CONTENTS

Targeting Maths Homework Year 4

ISBN: 978 1 925726 57 2

Published by Pascal Press
PO Box 250
Glebe NSW 2037
www.pascalpress.com.au
contact@pascalpress.com.au

Author: Ann & Johnny Baker
Publisher: Lynn Dickinson
Editor: Ruth Schultz
Cover and Text Designer: Janice Bowles

Typesetter: Ruth Schultz
Images & Illustrations: Dreamstime (unless otherwise indicated)

Printed by Wai Man Bookbinding (China) Ltd

Introduction

Targeting Maths Homework aims to build and reinforce Maths skills. This book supports the ACARA Australian Version 9·0 Curriculum for Year 4 and helps children to revise and consolidate what has been taught in the classroom. ACARA codes are shown on each unit and a chart explaining their content descriptions is on pages v and vi. The inside cover shows the topics in each unit.

The structure of this book

This book has 32 carefully graded four-page units.
The units are divided between:

- Number and Algebra
- Measurement and Space
- Statistics & Probability
- Problem Solving.

Assessment

Term Reviews follow Units 1–8, 9–16, 17–24 and 25–32 to test work covered during the term, and allow parents and carers to monitor their child's progress. Children are encouraged to mark each unit as it is completed and to colour in the traffic lights at the end of each segment. These results are then transferred to the Marking Grid. Parents and carers can see at a glance if their child is excelling or struggling!

NAPLAN*-style questions

Each unit has a set of three NAPLAN*-style questions that directly match the learning in the unit and mirror question styles that frequently occur in NAPLAN. This is to ensure that children's familiarity and confidence with NAPLAN*-style questions is consistently developed.

Green = Excellent — 2 or fewer questions incorrect
Orange = Passing — 50% or more questions answered correctly
Red = Struggling — fewer than 50% correct and needs help

SCORE	/10	0-4	5-8	9-10

How to Use This Book

The activities in this book are specifically designed to be used at home with minimal resources and support. Helpful explanations of key concepts and skills are provided throughout the book to help understand the tasks. Useful examples of how to do the activities are provided.

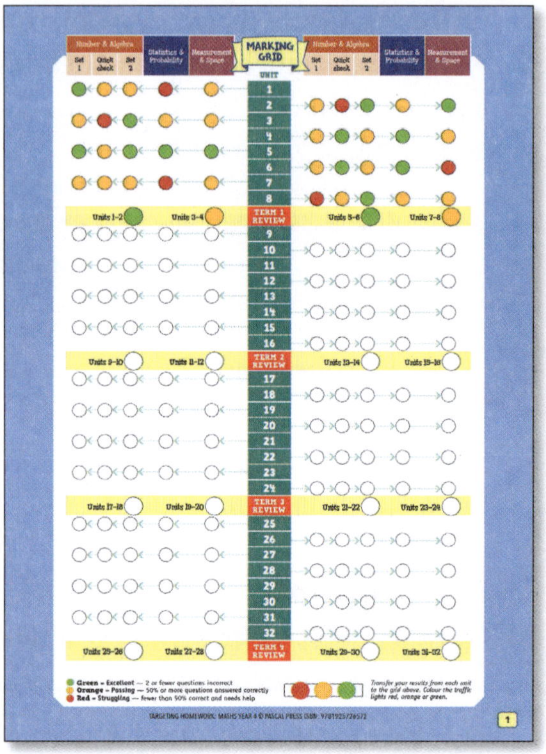

NOTE: The Problem of the Week sections do not appear on the Marking Grid as they often have multiple or subjective answers that cannot be easily scored.

Regular practice of key concepts and skills will support the work your child does in school and will enable you to monitor their progress throughout the year. It is recommended that children complete 8 units per school term (one a week) and then the Term Review. Every unit has a Traffic Light scoreboard at the end of each section.

You or your child should mark each completed unit and then colour the traffic light that corresponds to the number of correct questions. This process will enable you to see at a glance how your child is progressing and to identify weak spots. The results should be recorded at the end of each term on the Marking Grid on page 1.

The Term Review results are important for tracking progress and identifying any improvements in performance. If you find that certain questions are repeatedly causing difficulties and errors, then there is a good reason to discuss this with your child's teacher and arrange for extra instruction in that problem area.

Quick Maths Facts

The Quick Maths Facts on page 146 include multiplication tables up to 12 times, measurement equivalents, simple decimal–fraction–percentage conversions, a visual reminder of place value into the millions and the 'Days of the Month' rhyme. An accurate ruler has also been provided. Copy and laminate this page so that it is always available as a handy reference for your child.

Answers

The answer section on pages 147–162 can be removed, stapled together and kept somewhere safe. Use it to check answers when your child has completed each unit. Encourage your child to colour in the Traffic Light boxes when the answers have been calculated.

Australian Curriculum Correlations V9

Year 4 Maths		Number & Algebra	Statistics & Probability	Measurement & Space
ACARA CODE	**CONTENT DESCRIPTION**	**UNITS**	**UNITS**	**UNITS**
NUMBER				
AC9M4N01	Recognise and extend the application of place value to tenths and hundredths and use the conventions of decimal notation to name and represent decimals	1, 5, 9, 10, 11, 13, 16, 17, 20, 23		13
AC9M4N02	Explain and use the properties of odd and even numbers	1, 7, 14, 27, 28		
AC9M4N03	Find equivalent representations of fractions using related denominators and make connections between fractions and decimal notation	2, 9, 15, 21, 23, 24, 30		
AC9M4N04	Count by fractions including mixed numerals; locate and represent these fractions as numbers on number lines	2, 5, 9, 17, 31		
AC9M4N05	Solve problems involving multiplying or dividing natural numbers by multiples and powers of 10 without a calculator, using the multiplicative relationship between the place value of digits	2, 12, 18, 19, 29		
AC9M4N06	Develop efficient strategies and use appropriate digital tools for solving problems involving addition and subtraction, and multiplication and division where there is no remainder	4, 8, 11, 13, 18, 22, 24, 27, 29		
AC9M4N07	Choose and use estimation and rounding to check and explain the reasonableness of calculations including the results of financial transactions	3, 4, 5, 17, 20, 22, 25, 26		
AC9M4N08	Use mathematical modelling to solve practical problems involving additive and multiplicative situations including financial contexts; formulate the problems using number sentences and choose efficient calculation strategies, using digital tools where appropriate; interpret and communicate solutions in terms of the situation	6, 12, 13, 16, 21, 27, 28, 29, 30, 31, 32		
AC9M4N09	Follow and create algorithms involving a sequence of steps and decisions that use addition or multiplication to generate sets of numbers; identify and describe any emerging patterns	7, 14, 15, 23, 26, 32		
ALGEBRA				
AC9M4A01	Find unknown values in numerical equations involving addition and subtraction, using the properties of numbers and operations	3, 4, 14, 19, 20, 21, 28		
AC9M4A02	Recall and demonstrate proficiency with multiplication facts up to 10 x 10 and related division facts; extend and apply facts to develop efficient mental strategies for computation with larger numbers without a calculator	6, 10, 13, 15, 18, 19		

Australian CURRICULUM

Year 4 Maths		Number & Algebra	Statistics & Probability	Measurement & Space
ACARA CODE	CONTENT DESCRIPTION	UNITS	UNITS	UNITS
MEASUREMENT				
AC9M4M01	Interpret unmarked and partial units when measuring and comparing attributes of length, mass, capacity, duration and temperature, using scaled and digital instruments and appropriate units			2, 9, 11, 13, 24, 28
AC9M4M02	Recognise ways of measuring and approximating the perimeter and area of shapes and enclosed spaces, using appropriate formal and informal units			3, 15, 18, 22, 23, 30, 31
AC9M4M03	Solve problems involving the duration of time including situations involving "am" and "pm" and conversions between units of time			5, 17, 20, 26, 29
AC9M4M04	Estimate and compare angles using angle names including acute, obtuse, straight angle, reflex and revolution, and recognise their relationship to a right angle			7, 20, 28
SPACE				
AC9M4SP01	Represent and approximate composite shapes and objects in the environment, using combinations of familiar shapes and objects			1, 8, 12, 19, 25, 32
AC9M4SP02	Create and interpret grid reference systems using grid references and directions to locate and describe positions and pathways			4, 14, 21, 27
AC9M4SP03	Recognise line and rotational symmetry of shapes and create symmetrical patterns and pictures, using dynamic geometric software where appropriate			6, 16, 19, 30
STATISTICS				
AC9M4ST01	Acquire data for categorical and discrete numerical variables to address a question of interest or purpose using digital tools; represent data using many-to-one pictographs, column graphs and other displays or visualisations; interpret and discuss the information that has been created		1, 4, 6, 9, 23, 31	
AC9M4ST02	Analyse the effectiveness of different displays or visualisations in illustrating and comparing data distributions, then discuss the shape of distributions and the variation in the data		7, 10, 12, 20, 26, 28	
AC9M4ST03	Conduct statistical investigations, collecting data through survey responses and other methods; record and display data using digital tools; interpret the data and communicate the results		3, 15, 17, 18, 27, 28, 30	
PROBABILITY				
AC9M4P01	Describe possible everyday events and the possible outcomes of chance experiments and order outcomes or events based on their likelihood of occurring; identify independent or dependent events		2, 11, 13, 14, 16, 21, 22, 25, 29	
AC9M4P02	Conduct repeated chance experiments to observe relationships between outcomes; identify and describe the variation in results		2, 5, 8, 19, 32	

Australian CURRICULUM

TARGETING MATHS HOMEWORK: YEAR 4 © PASCAL PRESS ISBN: 9781925726572

MARKING GRID

Number & Algebra			Statistics & Probability	Measurement & Space	UNIT	Number & Algebra			Statistics & Probability	Measurement & Space
Set 1	Quick check	Set 2				Set 1	Quick check	Set 2		
					1					
					2					
					3					
					4					
					5					
					6					
					7					
					8					
Units 1–2			Units 3–4		TERM 1 REVIEW	Units 5–6			Units 7–8	
					9					
					10					
					11					
					12					
					13					
					14					
					15					
					16					
Units 9–10			Units 11–12		TERM 2 REVIEW	Units 13–14			Units 15–16	
					17					
					18					
					19					
					20					
					21					
					22					
					23					
					24					
Units 17–18			Units 19–20		TERM 3 REVIEW	Units 21–22			Units 23–24	
					25					
					26					
					27					
					28					
					29					
					30					
					31					
					32					
Units 25–26			Units 27–28		TERM 4 REVIEW	Units 29–30			Units 31–32	

Green = Excellent — 2 or fewer questions incorrect
Orange = Passing — 50% or more questions answered correctly
Red = Struggling — fewer than 50% correct and needs help

Transfer your results from each unit to the grid above. Colour the traffic lights red, orange or green.

TARGETING MATHS HOMEWORK: YEAR 4 © PASCAL PRESS ISBN: 9781925726572

Number & Algebra

AC9M4N01

Introducing Decimals

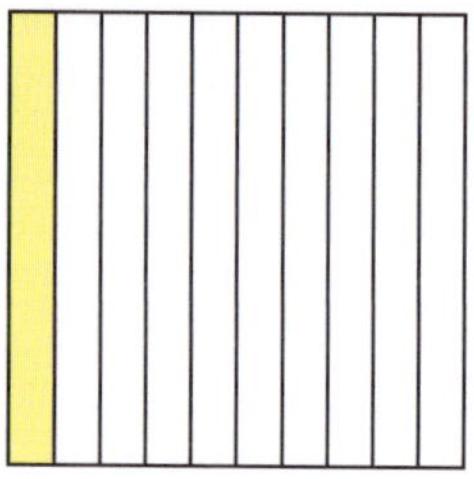

This decimal grid is divided into 10 equal strips.

Each strip is one-tenth of the whole grid.

The yellow strip is $\frac{1}{10}$ of the grid.

One-tenth can be written as a decimal: 0·1.

0·1

0 means no whole number.

1 means one-tenth.

The decimal point separates the whole number from the decimal part.

Label these diagrams as a fraction of the grid and as a decimal.

1. 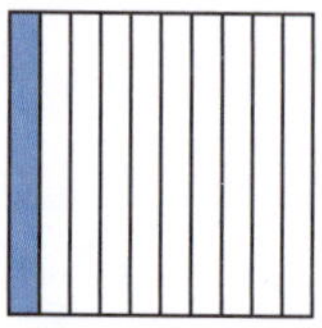 $\frac{1}{10}$ = 0·1
2. 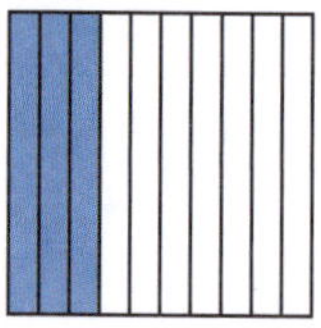 $\frac{\square}{10}$ = ___
3. 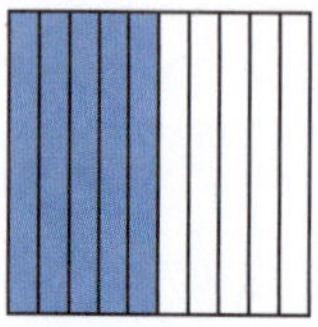 ___ = ___
4. 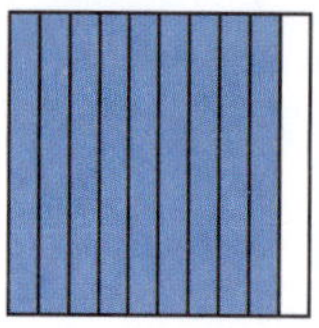 ___ = ___
5.  ___ = ___
6.  ___ = ___

Label these diagrams as fractions of the grids and as decimals.

7. 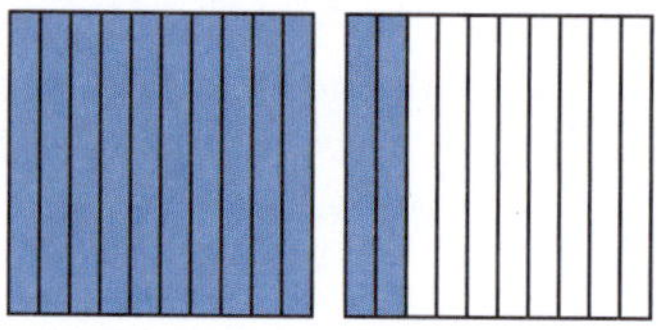 $\frac{12}{10}$ = 1·2
8. 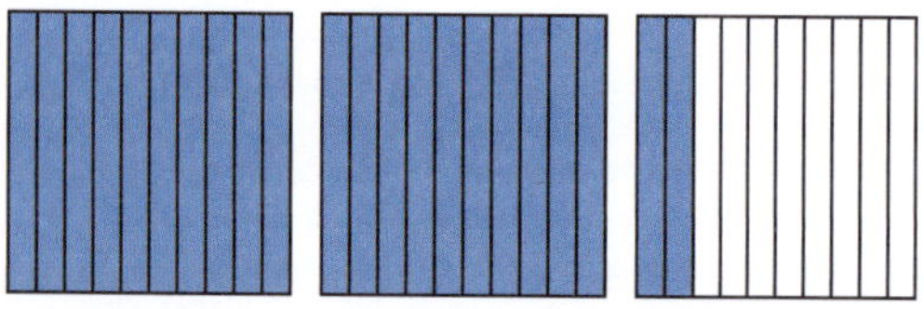 $\frac{\square}{10}$ = ___
9. 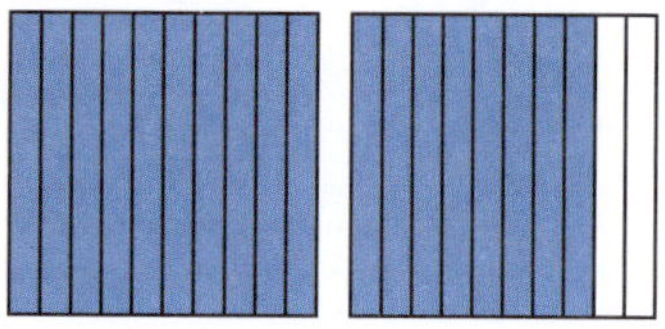 ___ = ___
10. 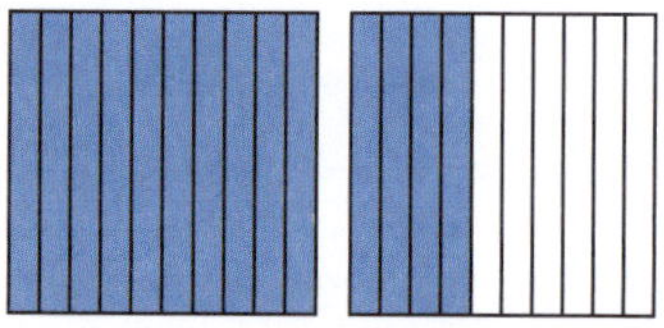 ___ = ___

SCORE /10 0-4 5-8 9-10

Quick Check

1. $\frac{4}{10}$ = ___
2. ___ = 0·7
3. $\frac{5}{10}$ = ___
4. $\frac{3}{10}$ = ___
5. ___ = 0·9
6. $\frac{14}{10}$ = ___
7. ___ = 1·3
8. $\frac{17}{10}$ = ___
9. ___ = 2·1
10. $\frac{28}{10}$ = ___

SCORE /10 0-4 5-8 9-10

TARGETING MATHS HOMEWORK: YEAR 4 © PASCAL PRESS ISBN: 9781925726572

Number & Algebra

AC9M4N02

Odd and Even Numbers

A number that can be divided into two parts that are equal in size is an **even number**.
If the parts are not the same size, the number is an **odd number**.

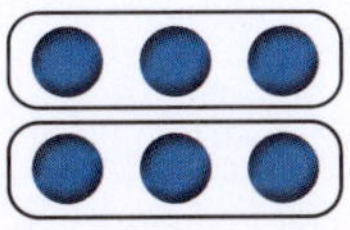

A group of 6 counters can be divided into two groups of 3 counters.
6 is an **even number**.

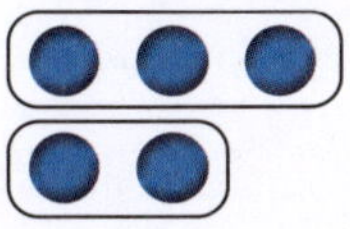

A group of 5 counters can be divided into two parts: a group of 3 and a group of 2.
5 is an **odd number**.

Divide the collections of counters into two groups to see if the number of counters is even or odd.

① 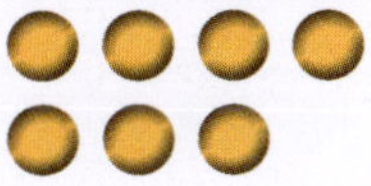

7 is ________.

② 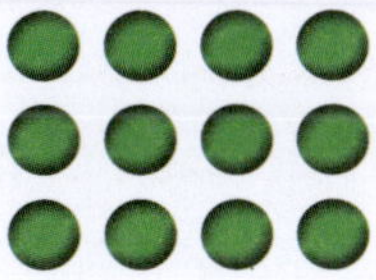

12 is ________.

③ 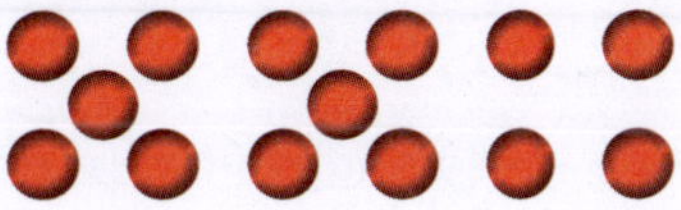

____ is ________.

All numbers in the 2s counting pattern are even.
Odd numbers come between two even numbers.

④ **Continue the sequences of odd and even numbers.**

	2		4		6												
1		3		5													

Are these numbers odd or even?

⑤ 16 is ________.

⑥ 17 is ________.

⑦ 23 is ________.

⑧ 70 is ________.

⑨ 32 is ________.

⑩ 100 is ________.

SCORE /10 0-4 5-8 9-10

Statistics & Probability

AC9M4ST01

Birthdays by the Seasons

Month	Number of birthdays
Jan.	26
Feb.	18
Mar.	34
Apr.	19
May	27
Jun.	21
Jul.	32
Aug.	16
Sep.	27
Oct.	14
Nov.	15
Dec.	12

I think more people have their birthdays in summer than in winter, but Zoe disagrees.
I have surveyed my school to see who is right.
The table shows the data that I collected.

① How many summer birthdays in Claire's data? ________

② How many autumn birthdays in Claire's data? ________

③ How many winter birthdays in Claire's data? ________

④ How many spring birthdays in Claire's data? ________

⑤ Do you agree with Claire or Zoe? ________

Why? ________________

SCORE /5 0-2 3-4 5

AC9M4SP01

Joining Shapes

I have been making shape puzzles by joining these blocks to make a larger shape:

I make a shape and then ask my friends to show me which blocks I used to make the shape.

I used two triangle blocks to make this shape:

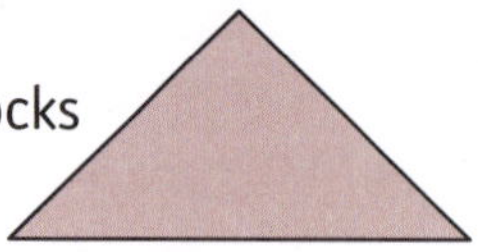

Here are the two triangle blocks I used:

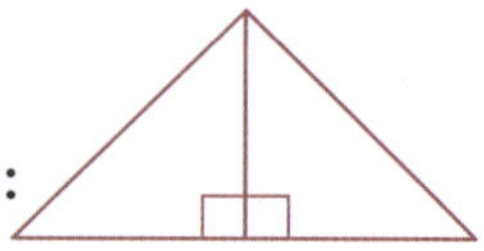

Show how Jake made these shapes and name the blocks he used to make them.

1. 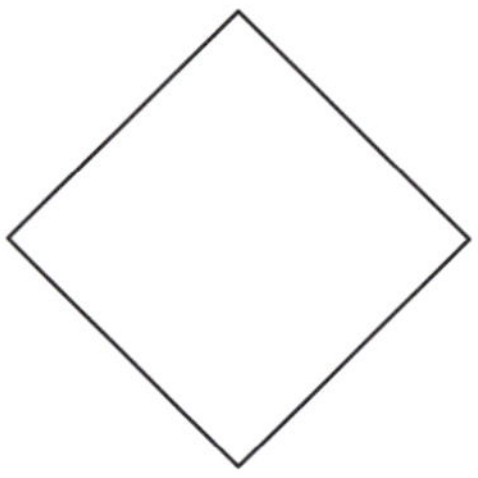

Shapes used:

____

____

2. 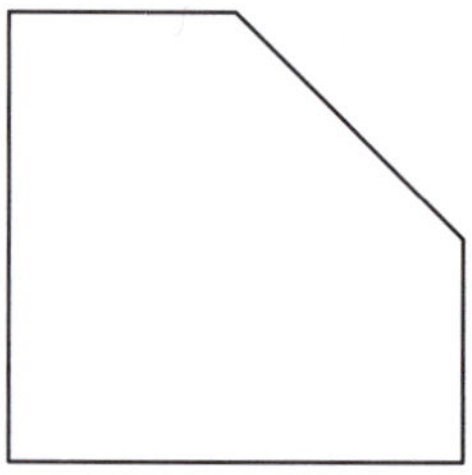

Shapes used:

____

____

____

3. 

Shapes used:

____

____

4. Jake used 5 blocks to make this shape:

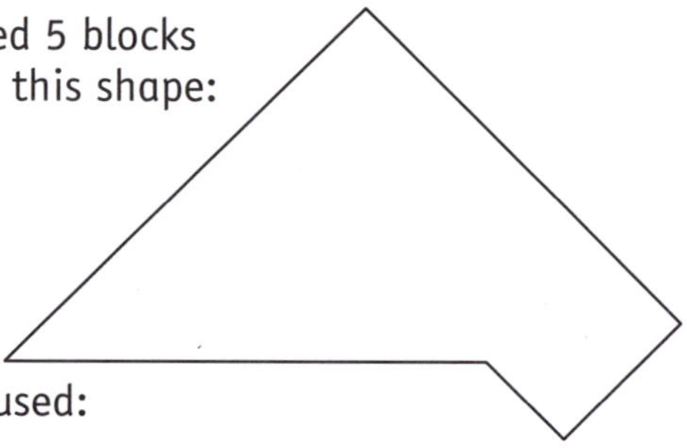

Shapes used:

____

____

SCORE /4 0-1 2-3 4

NAPLAN*-style Questions

1. Which of these grids has 0·6 shaded? Shade the bubble.

2. What is the next odd number in this sequence? 37, 39, ☐

3. The chart shows the number of goals scored this season.

Claire scored 6 goals.

How many did Dan score? ☐

SCORE /3 0-1 2 3

TARGETING MATHS HOMEWORK: YEAR 4 © PASCAL PRESS ISBN: 9781925726572

Problem of the Week

AC9M4ST02

The Greengrocer's Apple Graph

Last week the greengrocer kept a record of his apple sales. His apples are packed in boxes of 10 or 5.

But when he looked at his graph, he became confused because the blue line was meant to show the shape of the data.

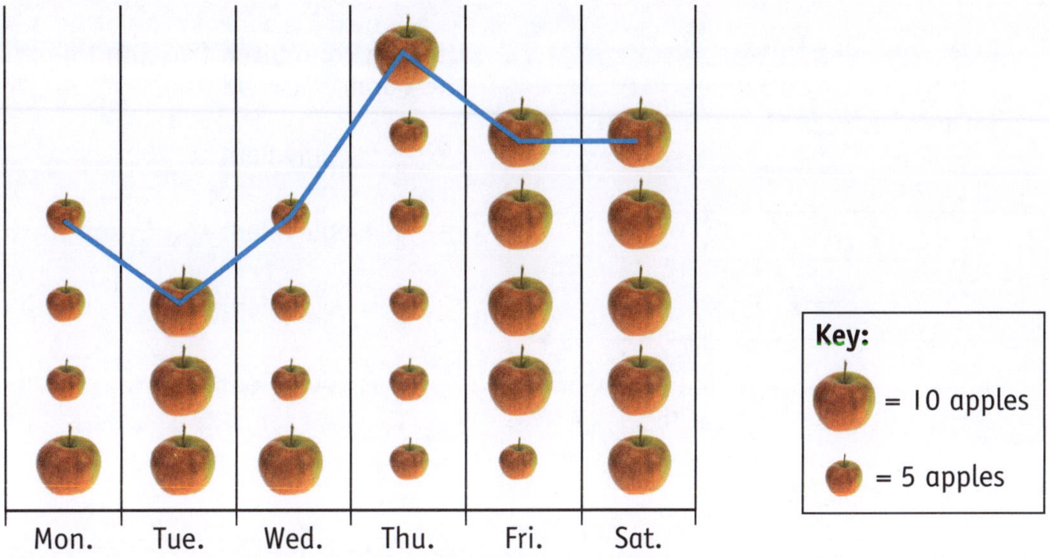

1. Without counting, which day does it look like the least number of apples was sold? ____________

2. Without counting, which day does it look like the most apples were sold? ____________

3. Count to check your answers to questions 1 and 2.
   Now complete the following graph to show the data in a less confusing way.

Title: ______________________________

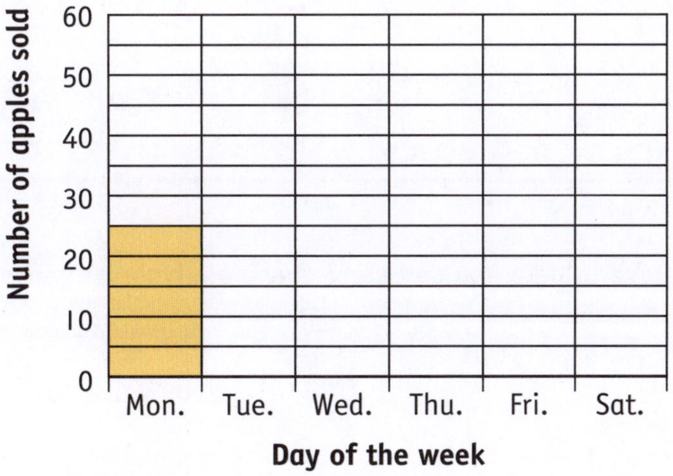

4. Is the shape of the new graph the same as the one that the grocer made? ________
   Why not? ______________________________

5. Is the new graph less confusing? ________
   Why is that? ______________________________

Number & Algebra

AC9M4N03, AC9M4N04

Equivalent Fractions

When we look at the fraction lines, we can see that $\frac{1}{2}$ is the same as (equivalent to) $\frac{2}{4}$.

0, $\frac{1}{2}$, 1

0, $\frac{1}{4}$, $\frac{2}{4}$, $\frac{3}{4}$, 1

Complete these fractions lines.

1. 0, $\frac{\square}{\square}$, 1 or $\frac{2}{2}$
2. 0, $\frac{\square}{\square}$, $\frac{\square}{\square}$, $\frac{\square}{\square}$, 1 or $\frac{4}{4}$
3. 0, $\frac{\square}{\square}$, $\frac{\square}{\square}$, $\frac{\square}{\square}$, $\frac{\square}{\square}$, $\frac{\square}{\square}$, $\frac{\square}{\square}$, $\frac{\square}{\square}$, 1 or $\frac{8}{8}$

Use the completed fraction lines to fill in the gaps.

4. $\frac{1}{4}$ is equivalent to —.
5. $\frac{4}{8}$ is equivalent to — and to —.
6. $\frac{6}{8}$ is equivalent to —.
7. $\frac{8}{8}$ is equivalent to —.

Complete these fraction lines.

8. 0, $\frac{1}{3}$, $\frac{\square}{\square}$, 1
9. 0, $\frac{1}{6}$, $\frac{\square}{\square}$, $\frac{\square}{\square}$, $\frac{\square}{\square}$, $\frac{\square}{\square}$, 1
10. 0, $\frac{\square}{\square}$, 1
11. 0, $\frac{1}{10}$, $\frac{\square}{\square}$, $\frac{\square}{\square}$, 1

Write two fractions that are equivalent to $\frac{1}{2}$.

12. $\frac{\square}{\square}$
13. $\frac{\square}{\square}$

SCORE /13 0-6 7-11 12-13

Quick Check

Use the fraction lines above to help you complete these equivalence statements.

1. $\frac{1}{4}$ is equivalent to —.
2. $\frac{3}{6}$ is equivalent to —.
3. $\frac{5}{10}$ is equivalent to —.
4. $\frac{1}{5}$ is equivalent to —.
5. $\frac{1}{3}$ is equivalent to —.
6. $\frac{3}{5}$ is equivalent to —.
7. $\frac{10}{10}$ is equivalent to —.
8. $\frac{4}{6}$ is equivalent to —.
9. $\frac{6}{10}$ is equivalent to —.
10. $\frac{3}{4}$ is equivalent to —.

SCORE /10 0-4 5-8 9-10

TARGETING MATHS HOMEWORK: YEAR 4 © PASCAL PRESS ISBN: 9781925726572

Number & Algebra

AC9M4N05

Multiplying by 10 and 100

Look at how the positions of the numerals change when you multiply by 10 or 100.

Start				× 10	Result			
Th	H	T	U		Th	H	T	U
			1	⇨			1	0
		1	0	⇨		1	0	0
	1	0	0	⇨	1	0	0	0
		1	5	⇨		1	5	0
		2	6	⇨		2	6	0
	1	2	3	⇨	1	2	3	0

Start				× 100	Result			
Th	H	T	U		Th	H	T	U
			1	⇨		1	0	0
		1	0	⇨	1	0	0	0
	1	0	0	⇨	10	0	0	0
		1	5	⇨	1	5	0	0
		2	6	⇨	2	6	0	0
	1	2	3	⇨	12	3	0	0

What will these numbers be after multiplying by 10?

1. 23 ⇨ ________
2. 106 ⇨ ________
3. 36 ⇨ ________
4. 127 ⇨ ________
5. 145 ⇨ ________
6. 350 ⇨ ________

What will these numbers be after multiplying by 100?

7. 36 ⇨ ____________
8. 145 ⇨ ____________
9. 45 ⇨ ____________
10. 360 ⇨ ____________

SCORE /10 0-4 5-8 9-10

Statistics & Probability

AC9M4P01, AC9M4P02

Just by Chance

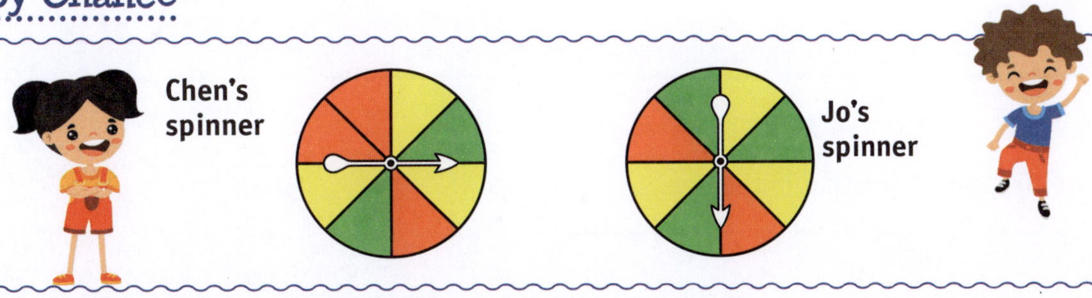

Use Chen's and Jo's spinners to answer the following questions.

1. Who is more likely to spin red? ____________
2. Who is less likely to spin green? ____________
3. After 8 spins, how many reds do you think Chen will have spun? ________
4. After 8 spins, how many reds do you think Jo will have spun? ________
5. What is the chance that Jo will spin green?

◯ Impossible ◯ Not likely ◯ Likely ◯ Very likely ◯ Certain

SCORE /5 0-2 3-4 5

AC9M4M01

Telling the Time

These clocks only have numerals for the quarter hours: 12, 3, 6 and 9. But you can still tell the times that they show. Write the times in digital form.

1. 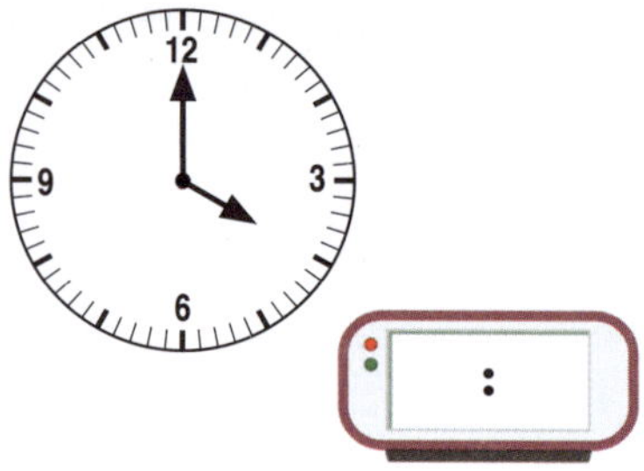

3. 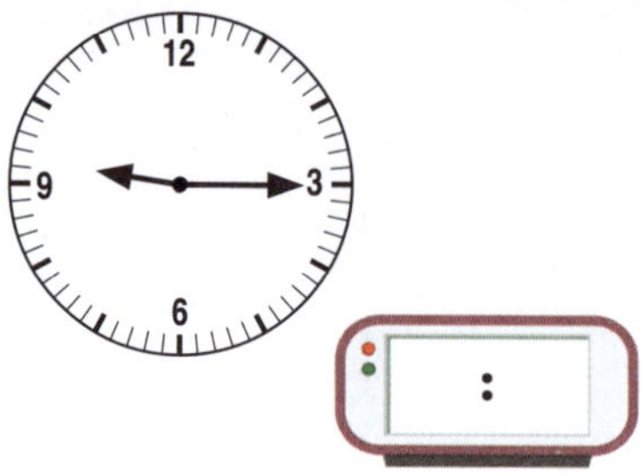

5. 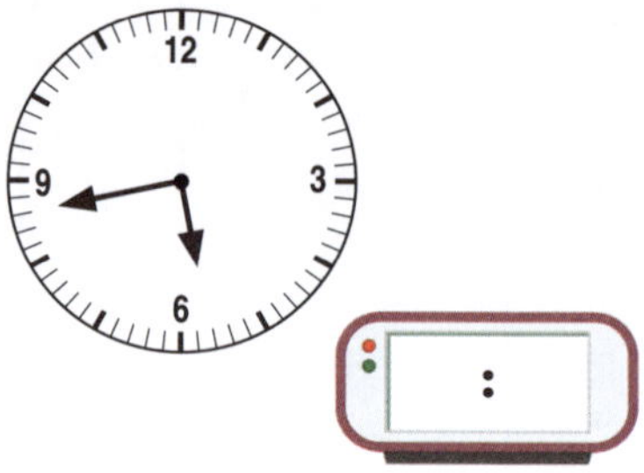

2. 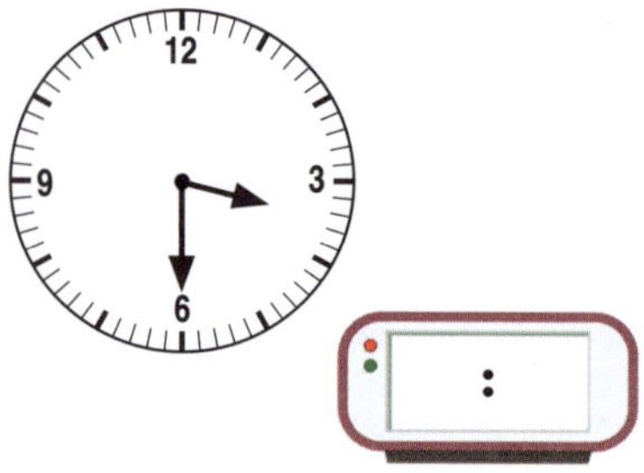

4. 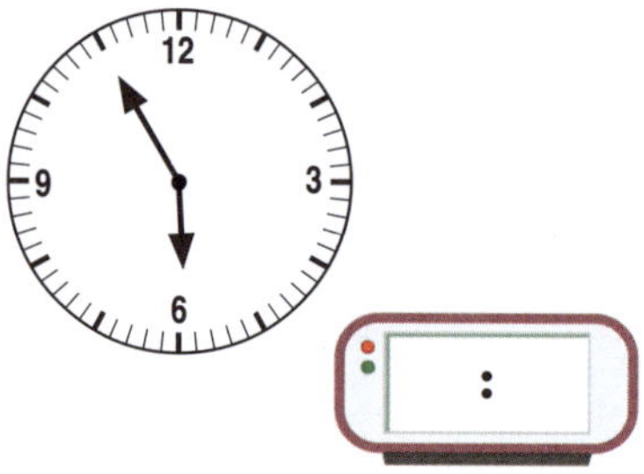

6. 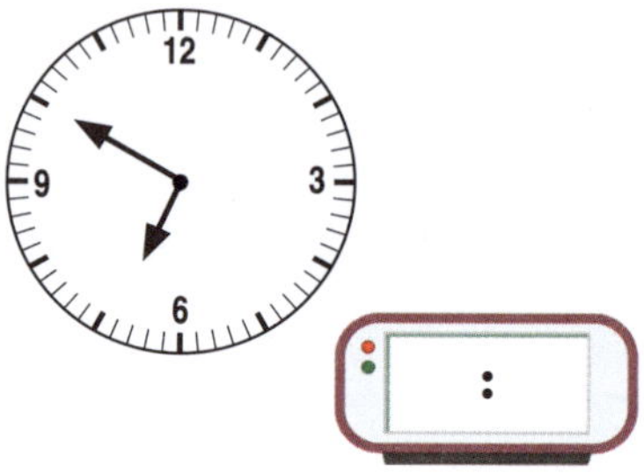

How much liquid is in these jugs? Answer in millilitres.

7. ________ mL

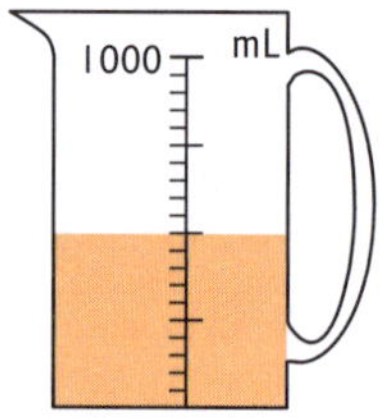

8. ________ mL

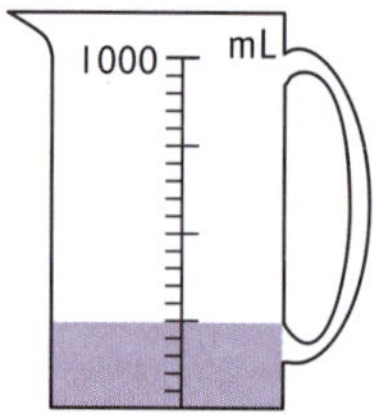

9. ________ mL

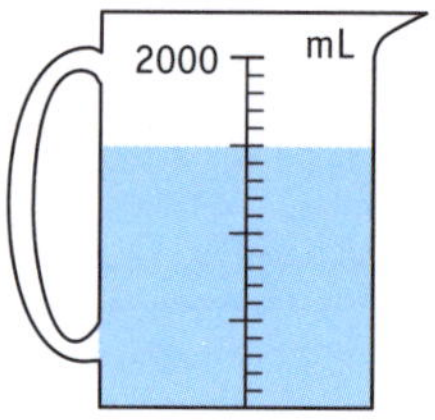

SCORE /9 0-4 5-7 8-9

NAPLAN*-style Questions

1. Write the correct fraction in the box.

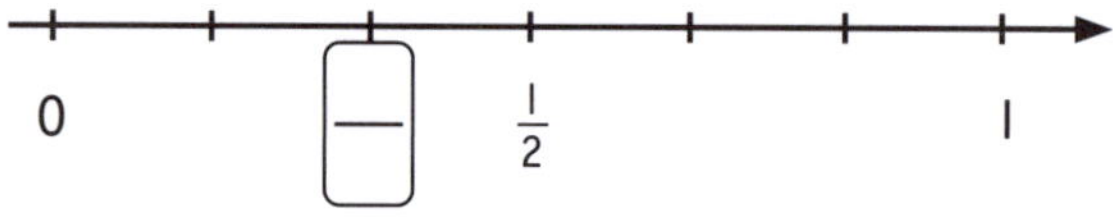

2. Claire spent $13. Jake spent 10 times as much. How much more than Claire did Jake spend?

◯ $13 ◯ $10 ◯ $130 ◯ $117

3. Which of these spinners is most likely to land on red?

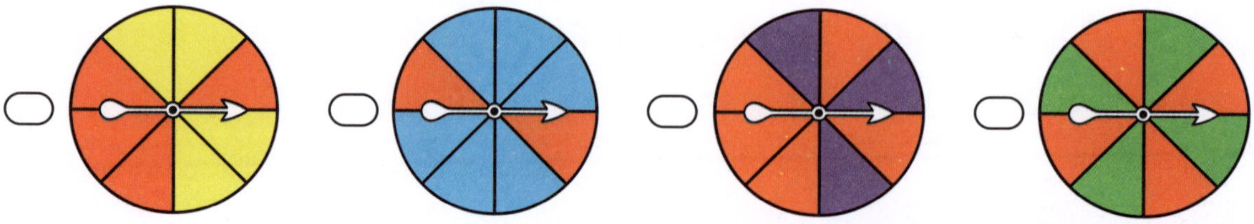

SCORE /3 0-1 2 3

TARGETING MATHS HOMEWORK: YEAR 4 © PASCAL PRESS ISBN: 9781925726572

Problem of the Week

A Scissors–Paper–Rock Investigation

This is a traditional game for two players.

- Each round, the players tap a closed hand on the table three times as they say: "Scissors, Paper, Rock, Go" in unison.
- On the word "Go", each player makes one of the above hand shapes.
- The outcome of the round is shown in the table.
- For example, if Player 1 played Scissors and Player 2 played Paper, then Player 1 is the winner.
- To remember the outcomes, remember that scissors cut paper, paper wraps rock, and rock blunts scissors.
  If both players make the same shape, the result is a draw.

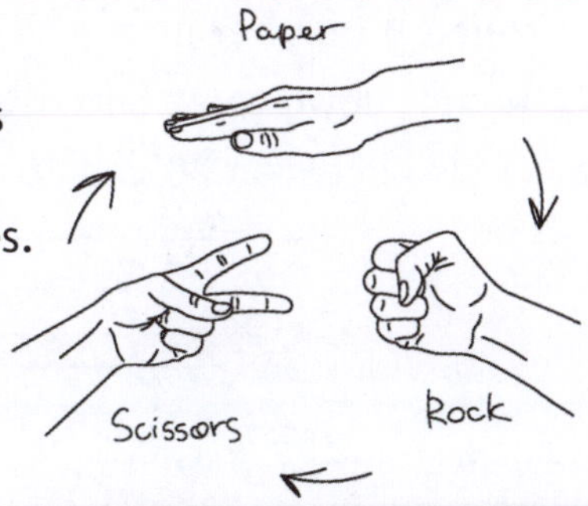

		Player 2		
		Scissors	**Paper**	**Rock**
Player 1	**Scissors**	Draw	Win	Lose
	Paper	Lose	Draw	Win
	Rock	Win	Lose	Draw

Is there a winning strategy?

To find out, you and a partner need to play 30 rounds of the game and record the results.

First, rate the three moves as having the best, equal or least chance of winning the round.

Scissors	Paper	Rock

In the table below, use tallies to record which moves are made as you play.

		Player 2		
		Scissors	Paper	Rock
Player 1	Scissors			
	Paper			
	Rock			

For example, suppose that you are Player 1 and you make Scissors.
Your partner is Player 2 and they make Paper.
Because scissors cut paper, you win. Make a tally mark in the middle box of the top row.

Play 30 rounds. This doesn't have to be all in one session or even with just one partner.
Then complete this table of wins, draws and losses.

You played	Wins	Draws	Losses
Scissors			
Paper			
Rock			

1. What conclusions can you make from the data that you collected?

____

____

2. Do you think that Scissors–Paper–Rock is a fair game? _______

Rounding and Estimating

To **round a number** means changing it to the nearest 10.

22 rounds down to 20 because is closer to 20 than to 30.

27 is also between 20 and 30. It is closer to 30 than 20 and so rounds up to 30.

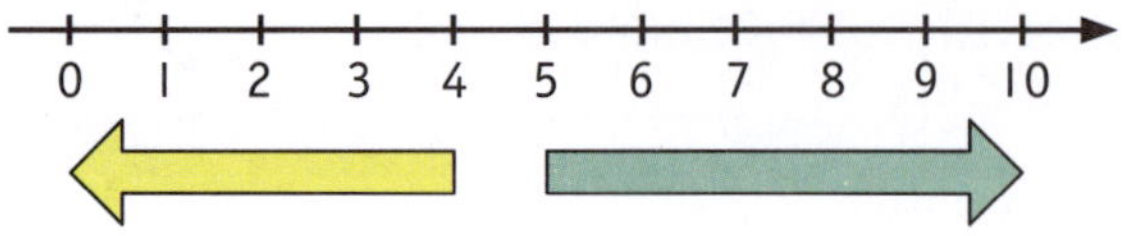

Numbers that have 1, 2, 3 or 4 in the ones position round down to the 10 below.
Numbers that have 6, 7, 8, or 9 are closer to the 10 above and round up to the 10 above.
5 is in the middle but always rounds up to the 10 above.

Round to the nearest 10. The first one is done for you.

1. 26 ➤ 30
2. 39 ➤ ____
3. 25 ➤ ____
4. 71 ➤ ____

You can also round to the nearest 100.

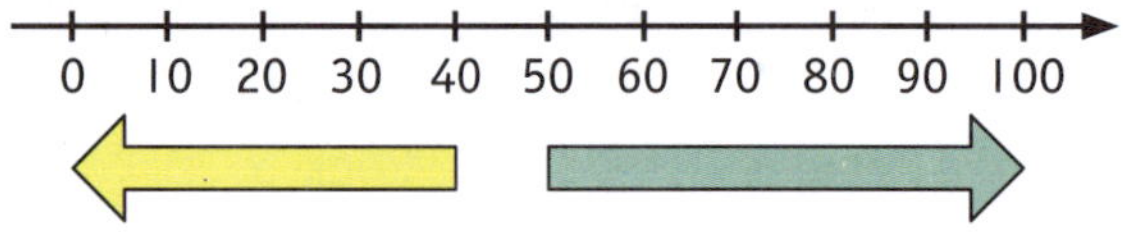

Round to the nearest 100. The first one is done for you.

5. 123 ➤ 100
6. 290 ➤ _____
7. 351 ➤ _____
8. 410 ➤ _____

Round to the nearest 10 and to the nearest 100.

Example: 135 rounds up to 140 when rounding to the nearest 10
and 135 rounds down to 100 when rounding to the nearest 100.

9. 374 rounds ________ to _____ when rounding to the nearest _____.
   374 rounds ________ to _____ when rounding to the nearest _____.
10. 415 rounds ________ to _____ when rounding to the nearest _____.
    415 rounds ________ to _____ when rounding to the nearest _____.

SCORE /10 0-4 5-8 9-10

Quick Check

Round to the nearest 10.

1. 56 ➤ ______
2. 32 ➤ ______
3. 102 ➤ ______
4. 148 ➤ ______
5. 199 ➤ ______

Round to the nearest 100.

6. 632 ➤ ______
7. 428 ➤ ______
8. 355 ➤ ______
9. 899 ➤ ______
10. 320 ➤ ______

SCORE /10 0-4 5-8 9-10

Number & Algebra

AC9M4A01

Addition and Subtraction with Unknowns

In these problems, one number is missing or unknown.
It could be the **answer**, **start** number or **change** number.
You can use a bar model to find the missing number.

The Problem: There were six apples in the bowl and now there are 10 apples in the bowl.
How many apples have been added to the bowl?

The problem can be shown on a bar model.

10	
6	4

Complete the bar models.

1. There were 10 apples in the bowl after Kai added two apples.
How many apples were in the bowl to start with?

10	
	2

2. There are four green apples and eight red apples in the bowl.
How many apples altogether?

3. After Dan had eaten some of the apples, there were six left in the bowl, which had nine apples to start with.
How many apples did Dan eat?

9	

SCORE /3 0-1 2 3

Statistics & Probability

AC9M4ST03

Cordial Survey

Title: ____________________

I surveyed 20 people about their favourite cordial.
I asked, "Which cordial do you like best: lemon, orange or mango?"

There were six people who liked lemon the best, eight people who liked orange the best and the rest liked mango the best.

1. Write a title for the graph.
2. Complete the key of the graph.
3. How many people liked mango the best?

______

4. Complete the graph.

Key
= ___ votes

SCORE /4 0-1 2-3 4

TARGETING MATHS HOMEWORK: YEAR 4 © PASCAL PRESS ISBN: 9781925726572

AC9M4M02

Perimeter

You can find the perimeter by writing the length of each side and then adding the lengths.

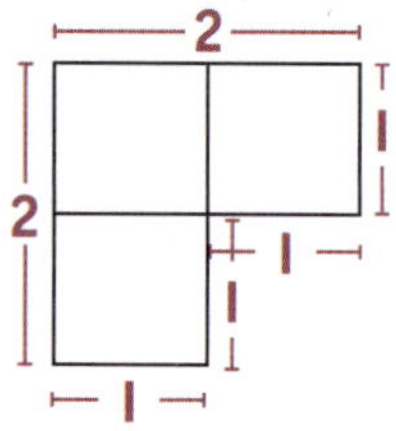

2 + 2 + 1 + 1 + 1 + 1 = 8
This shape has a perimeter of 8 cm.

The perimeter of a shape is the distance around the outside.

These shapes are made with 6 squares. Each square has a side length of 1 cm. Estimate and then check the perimeter of each shape.

① 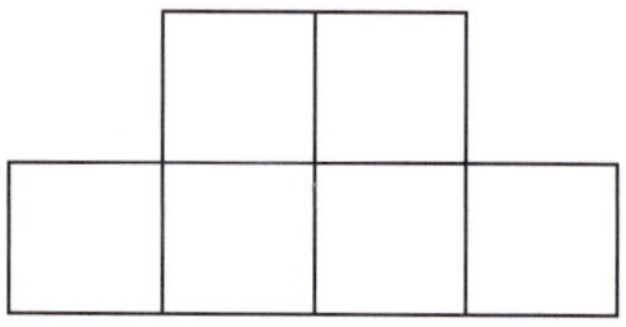

Estimate: ____ Check: ____

③ 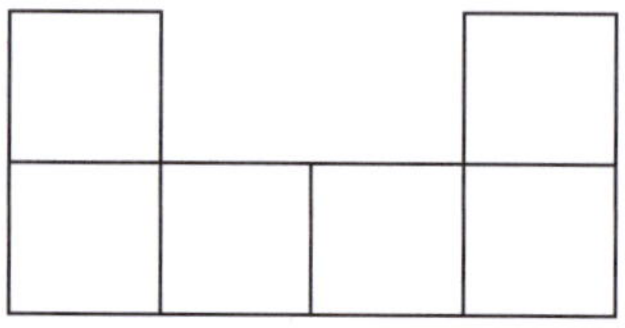

Estimate: ____ Check: ____

⑤ 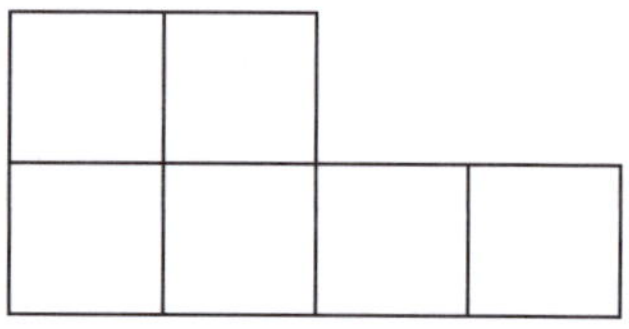

Estimate: ____ Check: ____

② 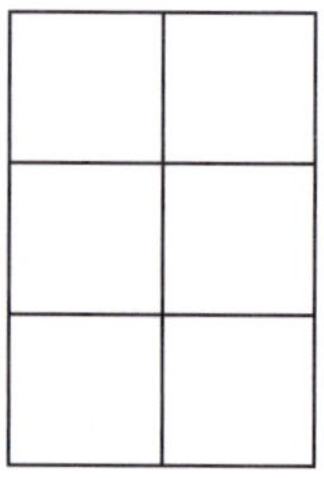

Estimate: ____ Check: ____

④ 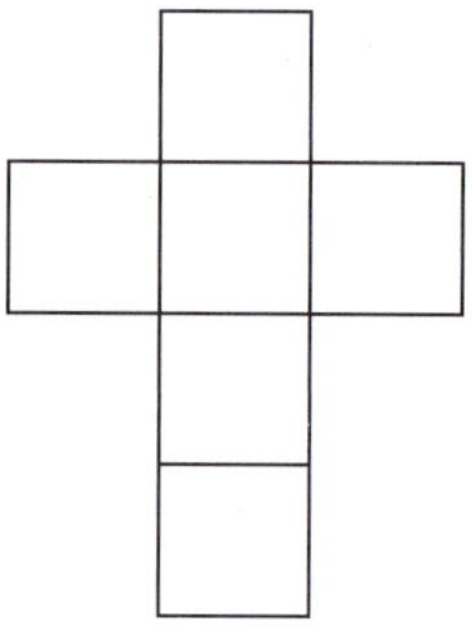

Estimate: ____ Check: ____

⑥ 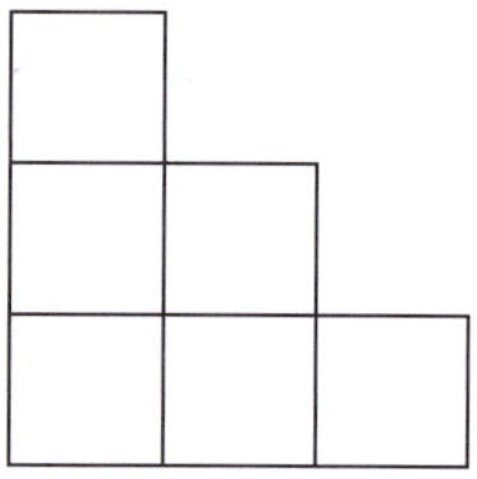

Estimate: ____ Check: ____

SCORE /6 0-2 3-4 5-6

NAPLAN*-style Questions

① Which of these numbers rounds up to the nearest 10 but down to the nearest 100?

◯ 144 ◯ 245 ◯ 376 ◯ 473

② Write the missing number in this number sentence.

26 + ☐ = 42

③ Jake wants to put a wrap around the cake that he has just made. The cake is in the shape of a rectangle that is 20 cm by 30 cm.
How long does the cake wrap need to be?

__________ cm.

SCORE /3 0-1 2 3

TARGETING MATHS HOMEWORK: YEAR 4 © PASCAL PRESS ISBN: 9781925726572

Problem of the Week

Perimeter Puzzles

Dan made a shape by joining eight 1 cm squares along an edge. To help find the perimeter of this shape, he labelled each side with its length, using different colours for the different lengths.

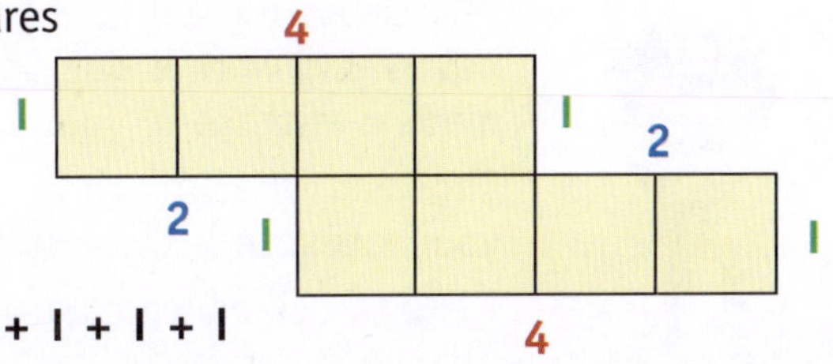

Dan saw that the addition **4 + 4 + 2 + 2 + 1 + 1 + 1 + 1** was very long and that it would be better as:
2 × 4 + 2 × 2 + 4 × 1 = 8 + 4 + 4 = 16.

Here are some more shapes that Dan made using eight squares. Use Dan's method to find the perimeter.

① 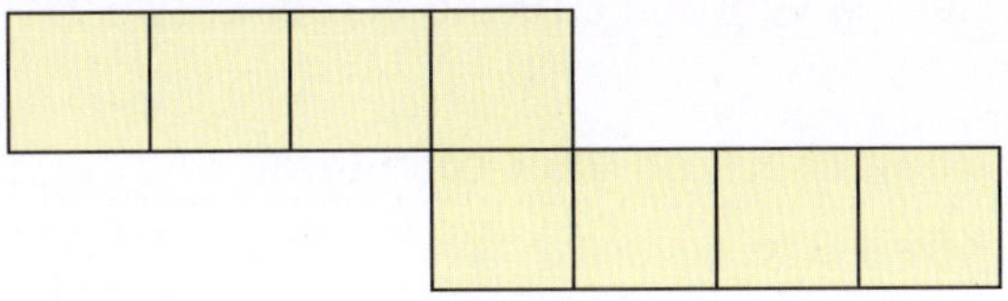

Perimeter = ______________________

____

② 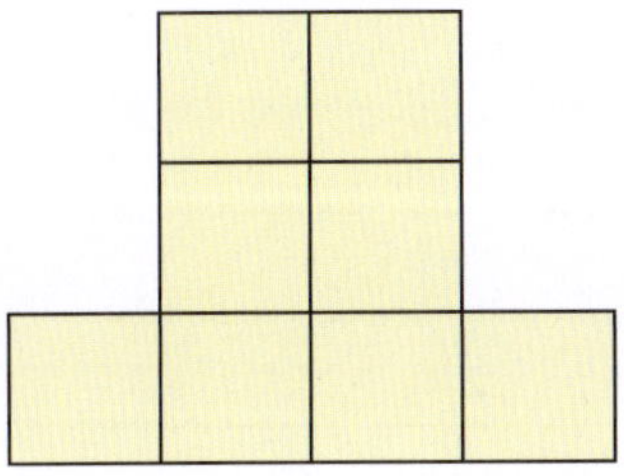

Perimeter = ______________________

____

③ 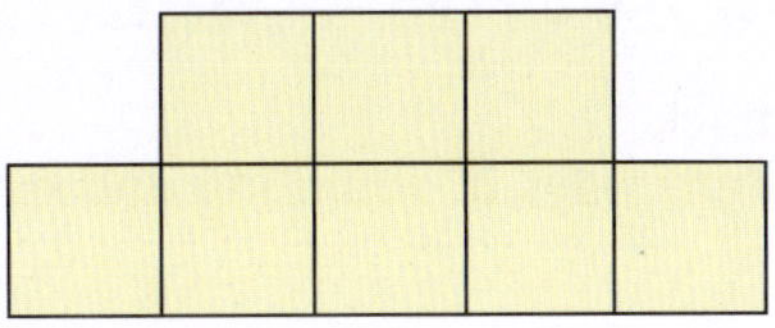

Perimeter = ______________________

____

④ 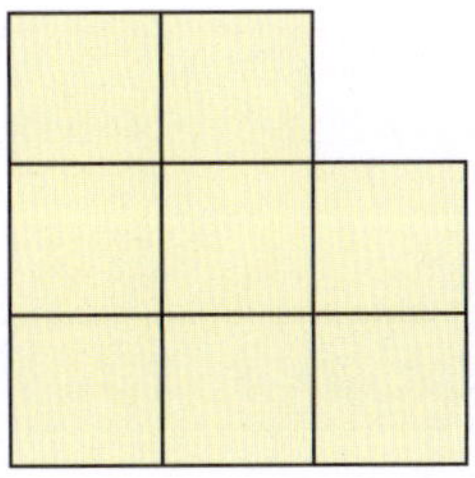

Perimeter = ______________________

____

Dan's Challenge

What different perimeters can you make using eight blocks and always joining edge to edge?
18 is the longest perimeter that can be made. What is the shortest perimeter?

Use this space to experiment.

AC9M4N06, AC9M4N07

Estimation and Addition

It's always a good idea to estimate the answer to an addition before you work it out. It is the best way to spot if you have made a mistake in your calculating.

Choosing what to round, up or down, will make sure that your estimate is close to the actual answer.

I like to use my calculator to see how close my estimate is.

Example: **Estimate the answer to 364 + 754.**

Rounding up to the nearest 100 gives **400 + 800 = 1200**.

Rounding both numbers to the nearest 10 gives **360 + 750 = 1110**.

Choose 1110 as the estimate.

Check: 364 + 756 = 1120. Yes, the chosen estimate is very close to the actual answer.

Estimate and check.
Show two different ways of rounding and estimating. Then check on your calculator.

1. 278 + 343
   Estimate 1 ____________
   Estimate 2 ____________
   Check ____________
2. 572 + 636
   Estimate 1 ____________
   Estimate 2 ____________
   Check ____________
3. 467 + 432
   Estimate 1 ____________
   Estimate 2 ____________
   Check ____________
4. 489 + 376
   Estimate 1 ____________
   Estimate 2 ____________
   Check ____________
5. 324 + 448
   Estimate 1 ____________
   Estimate 2 ____________
   Check ____________
6. 347 + 458
   Estimate 1 ____________
   Estimate 2 ____________
   Check ____________

SCORE /6 0-2 3-4 5-6

Quick Check

Make an estimate of the actual answer.

Round to the nearest 10

1. 335 + 894 ➤ ______ + ______ = ______
2. 278 + 247 ➤ ______ + ______ = ______
3. 389 + 433 ➤ ______ + ______ = ______
4. 486 + 225 ➤ ______ + ______ = ______
5. 375 + 225 ➤ ______ + ______ = ______

Round to the nearest 100

6. 865 + 428 ➤ ______ + ______ = ______
7. 739 + 628 ➤ ______ + ______ = ______
8. 676 + 525 ➤ ______ + ______ = ______
9. 876 + 929 ➤ ______ + ______ = ______
10. 1347 + 2458 ➤ ______ + ______ = ______

SCORE /10 0-4 5-8 9-10

TARGETING MATHS HOMEWORK: YEAR 4 © PASCAL PRESS ISBN: 9781925726572

Number & Algebra

AC9M4A01

Calculator Sequences

This is a problem that can be solved on a calculator.
Use this key sequence:

600 [−] 37 [=]

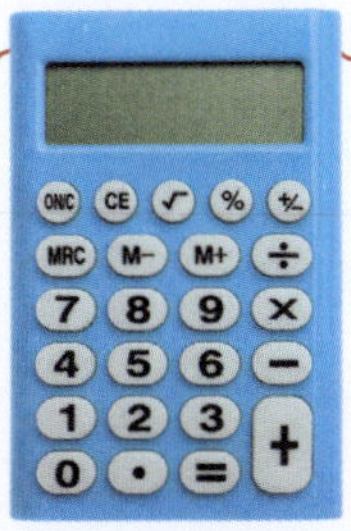

To write a calculator sequence, put these keys in a box:

[+] [−] [×] [÷] [=]

Write the calculator sequences and show the answers.

① It cost $126 for the jacket and $56 for the shirt. How much is that altogether?

___ [ ] ___ [=] ___

② There were 96 oranges for juicing. Now the juice bar has 18 oranges left. How many oranges did they juice?

___ [ ] ___ [ ] ___

③ The puppy has eaten 13 of the treats and there are 48 treats in a box. How many treats are left?

___ [ ] ___ [ ] ___

④ Kai had $28 and was paid to wash the car. Now he has $35. How much was he paid for washing the car?

___ [ ] ___ [ ] ___

⑤ After spending $32, Zoe has $129 left. How much did Zoe have to start with?

___ [ ] ___ [ ] ___

⑥ 24 blackbirds were caught under the cherry tree net, but 15 managed to escape. How many blackbirds were left to eat our cherries?

___ [ ] ___ [ ] ___

SCORE /6 0-2 3-4 5-6

Statistics & Probability

AC9M4ST01

At the Birdbath

Chen observed 120 birds using the birdbath last week.

She saw:

- 30 birds on Monday
- 15 birds on Tuesday and Wednesday
- 20 birds on Thursday
- 10 birds on Friday.

Chen saw the same number of birds on Saturday and Sunday.

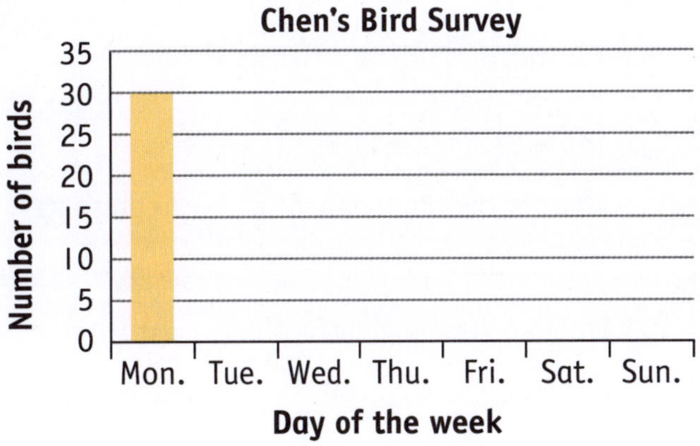

① Chen saw ____ birds during the school week.

② Chen saw ____ birds during the weekend.

③ Chen saw ____ birds on both Saturday and Sunday.

④ Complete the chart to show how many birds Chen saw.

SCORE /4 0-1 2-3 4

AC9M4SP02

Grid Coordinates

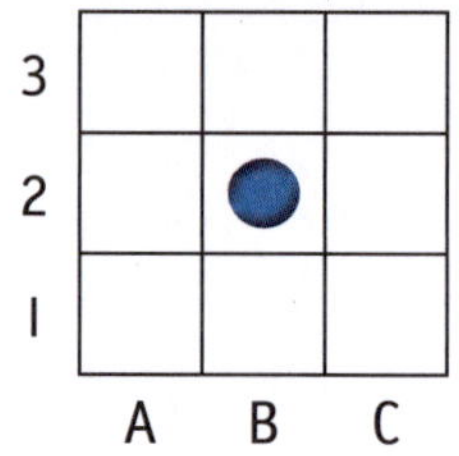

This is an alphanumeric grid that is used to locate items on a map.
Read the grid **across** then **up** to find the location.

On this grid, the counter is located at grid position **B2**.
Go across to B and then up to 2.

Where on the grid?

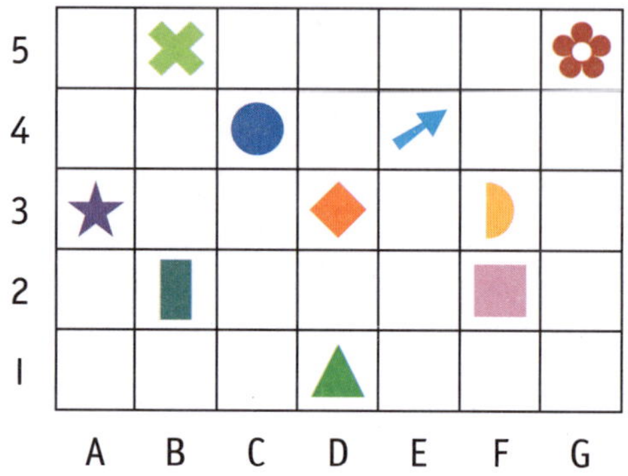

What are the grid coordinates of these objects on the grid?

1.  is at ______
2.  is at ______
3.  is at ______
4.  is at ______
5.  is at ______
6.  is at ______
7.  is at ______
8.  is at ______
9.  is at ______
10.  is at ______

SCORE /10 0-4 5-8 9-10

NAPLAN*-style Questions

1. $56 + $[ ] = $100
   What is the amount that [ ] represents?
   $ ________

2. Which of the options is the closest estimate for 389 + 427?
   ◯ 800 ◯ 700 ◯ 850 ◯ 900

3. What is the grid coordinate for the ?

    is at ______.

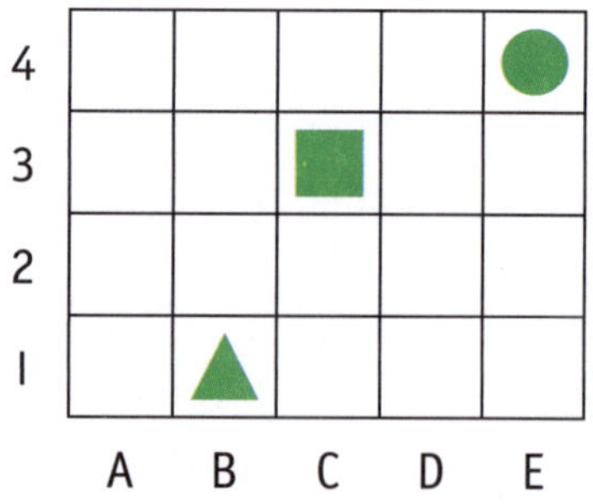

SCORE /3 0-1 2 3

TARGETING MATHS HOMEWORK: YEAR 4 © PASCAL PRESS ISBN: 9781925726572

Problem of the Week

AC9M4N07

How Close, How Far?

We have been investigating how close you get to the actual answer when you use estimation by rounding.

To do this:

- make up a four-number column addition, as shown below
- make an estimate of the answer by rounding each of the four numbers
- check the estimate with our calculator.

Over to you to have a go!

Rounding to the nearest 10

Make up some four-number addition strings, estimate the answers and check your estimates. The first one has been started for you.

① **Round and add**

36 ➤ 40
\+ 54 ➤ 50
\+ 52 ➤ ___
\+ 66 ➤ ___
= ___ ___

② **Round and add**

___ ➤ ___
\+ ___ ➤ ___
\+ ___ ➤ ___
\+ ___ ➤ ___
= ___ ___

③ **Round and add**

___ ➤ ___
\+ ___ ➤ ___
\+ ___ ➤ ___
\+ ___ ➤ ___
= ___ ___

Were the estimates close to the actual answer? ________

Why do you think that was? ________________________________

Rounding to the nearest 100

As above, make up some four-number addition strings, estimate the answers and check your estimates.

④ **Round and add**

536 ➤ 500
\+ 364 ➤ 400
\+ 566 ➤ _____
\+ 122 ➤ _____
= _____ _____

⑤ **Round and add**

_____ ➤ _____
\+ _____ ➤ _____
\+ _____ ➤ _____
\+ _____ ➤ _____
= _____ _____

⑥ **Round and add**

_____ ➤ _____
\+ _____ ➤ _____
\+ _____ ➤ _____
\+ _____ ➤ _____
= _____ _____

How close were the estimates?

⑦ Was rounding to 10 close to the actual answer?

◯ always ◯ sometimes ◯ never

⑧ Was rounding to 100 close to the actual answer?

◯ always ◯ sometimes ◯ never

AC9M4N01, AC9M4N04

Working with Tenths

My three spins were 0·3, $\frac{4}{10}$ and 0·7 and I coloured in the grid to match the spins.

Then I can find my total score: 1·4 or $1\frac{4}{10}$.

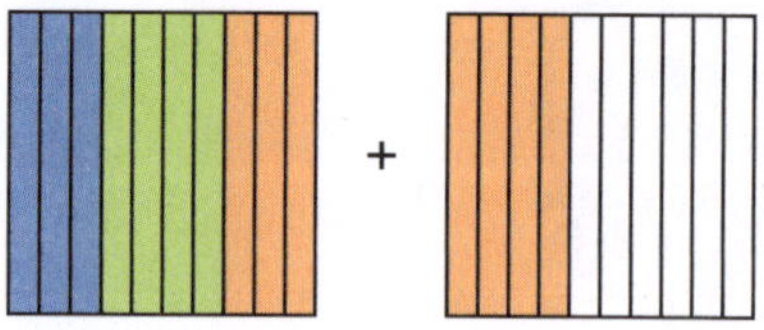

Colour the grids and work out the totals for these spins.

① Spins $\frac{4}{10}$, 0·7 and 0·4

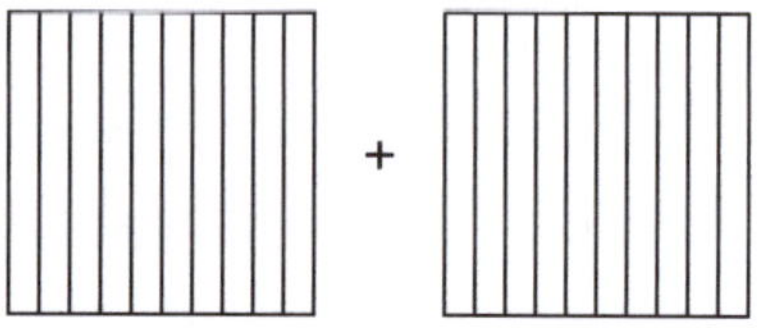

$\frac{4}{10}$ + 0·7 + 0·4 = ______

② Spins 0·4, $\frac{7}{10}$ and 0·7

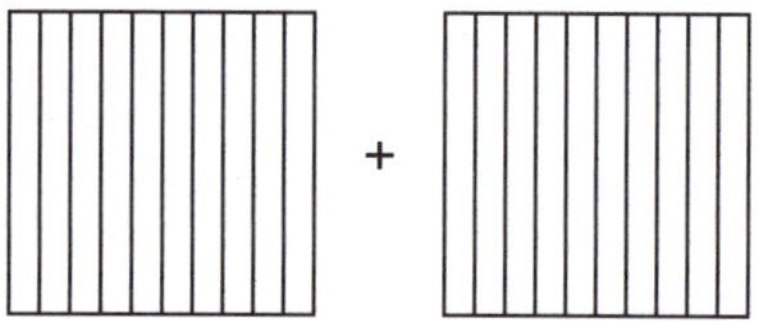

0·4 + $\frac{7}{10}$ + 0·7 = ______

③ Spins 0·3, $\frac{3}{10}$, and 0·7

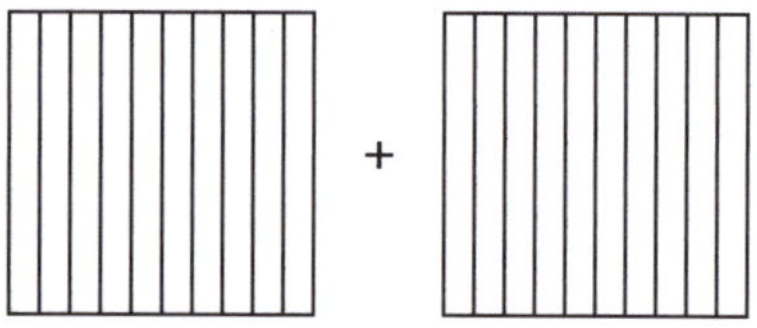

0·3 + $\frac{3}{10}$ + 0·7 = ______

④ Spins 0·4, $\frac{7}{10}$, and 0·3

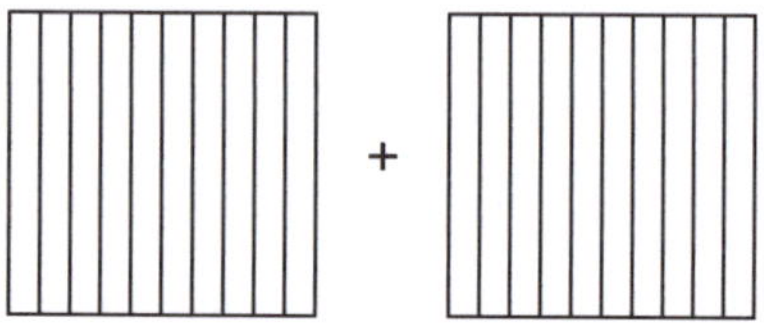

0·4 + $\frac{7}{10}$ + 0·3 = ______

⑤ Spins 0·7, $\frac{7}{10}$, and 0·3

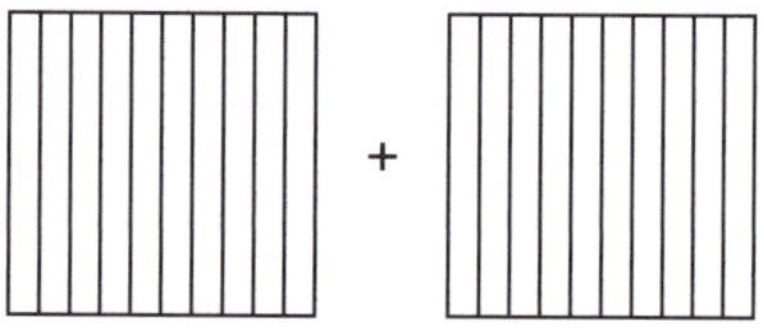

0·7 + $\frac{7}{10}$ + 0·3 = ______

SCORE /5 0-2 3-4 5

Quick Check

Use the decimal grid to help you work out the total of these fractions. Answer as a decimal and as a fraction.

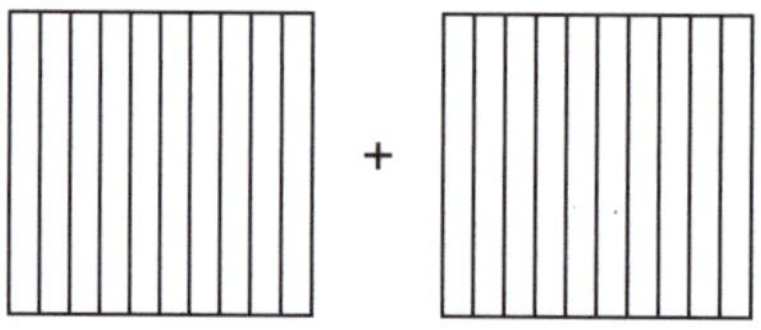

① $\frac{3}{10}$ + $\frac{1}{2}$ = 0·8 or $\frac{8}{10}$

② 0·7 + $\frac{6}{10}$ = __·__ or ——.

③ 0·4 + $\frac{6}{10}$ = __·__ or ——.

④ 0·5 + $\frac{6}{10}$ = __·__ or ——.

⑤ 0·6 + $\frac{5}{10}$ = __·__ or ——.

⑥ 0·5 + 0·5 + $\frac{5}{10}$ = __·__ or ——.

⑦ $\frac{6}{10}$ + 0·6 + $\frac{4}{10}$ = __·__ or ——.

⑧ 1·4 + $\frac{4}{10}$ + 0·6 = __·__ or ——.

⑨ 0·6 + 0·5 + $\frac{7}{10}$ = __·__ or ——.

⑩ 0·5 + 0·6 + $\frac{8}{10}$ = __·__ or ——.

SCORE /10 0-4 5-8 9-10

TARGETING MATHS HOMEWORK: YEAR 4 © PASCAL PRESS ISBN: 9781925726572

Estimation and Subtraction

In Unit 4, you used estimation based on rounding to the nearest 10 or 100 for **addition**.
We can round and estimate for **subtraction** too.
Jo was asked to make an estimate for 343 – 278. She chose to round to the nearest 10 first and then used her calculator to check.

Use rounding to the nearest 10 to estimate these subtractions and then check with a calculator.

① **Find** 478 – 153
Estimate 480 – 150 = _____
Check Actual answer = _____

② **Find** 637 – 495
Estimate _____ – _____ = _____
Check Actual answer = _____

③ **Find** 891 – 368
Estimate _____ – _____ = _____
Check Actual answer = _____

④ **Find** 547 – 352
Estimate _____ – _____ = _____
Check Actual answer = _____

⑤ **Find** 954 – 366
Estimate _____ – _____ = _____
Check Actual answer = _____

⑥ **Find** 1034 – 899
Estimate _____ – _____ = _____
Check Actual answer = _____

SCORE /6 0-2 3-4 5-6

Statistics & Probability

AC9M4P02

Counting Colours

Chen has an experiment she would like to share.
"I put one red pencil and one blue pencil in my pencil case and closed the zip. I shook the case and then closed my eyes while I opened up the pencil case and took out one of the pencils. I used that colour to make a tally of the results and repeated the experiment until I had done it 10 times. This is what I found."

R	卌 |
B	||||

① Before you try Chen's experiment yourself, make a prediction (estimate) about what you think will happen and then run the experiment to get the actual data.

Estimate		Actual	
R		R	
B		B	

② Repeat the experiment but this time put two red pencils and one blue pencil in the pencil case. Tally the data as before.

Estimate		Actual	
R		R	
B		B	

③ Was the actual result what you expected? _____

④ Why was that? ______________________________

SCORE /4 0-1 2-3 4

AC9M4M03

am and pm Times

The diagram shows the **am (ante meridiem)** and **pm (post meridiem)** parts of a 24-hour day.

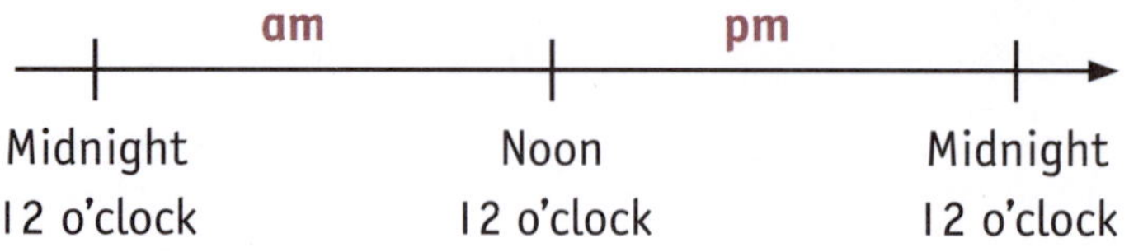

Use the letters to show where each event should be placed on the timeline.

1. **A** Start of the school day
2. **B** End of the school day
3. **C** Bedtime
4. **D** Breakfast time
5. **E** An hour before midnight

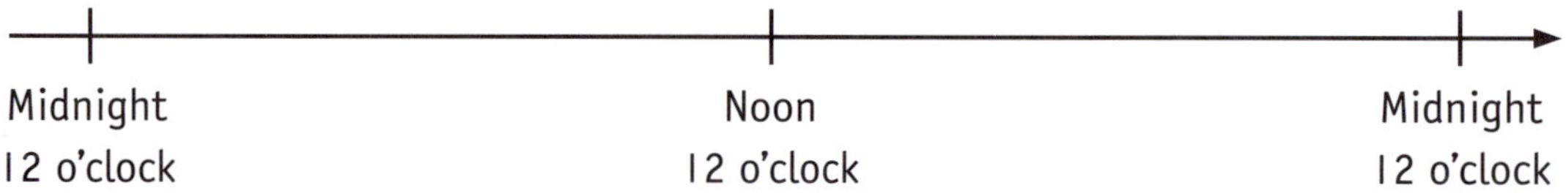

Label each event with the correct letter from the timeline.

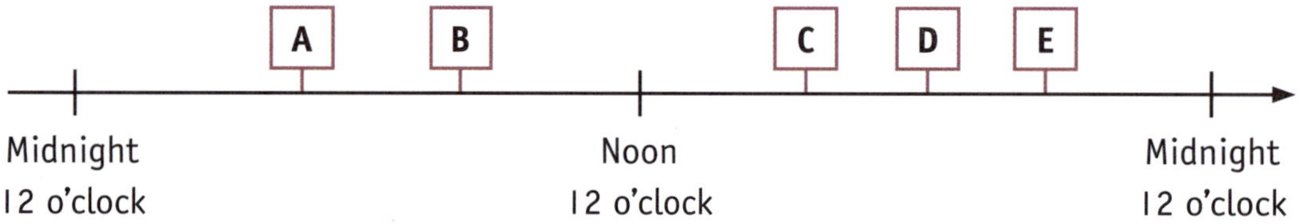

6. ☐ Morning tea
7. ☐ 8:00 pm
8. ☐ End of the school day
9. ☐ Dinner time
10. ☐ Time to get up

SCORE /10 (0-4) (5-8) (9-10)

NAPLAN*-style Questions

1. Which of these estimates is closest to this subtraction?

   175 – 86

   ◯ 70 ◯ 80 ◯ 90 ◯ 100

2. Kai rolls a regular dice six times. How likely is he to roll a 4?

   ◯ Impossible ◯ Unlikely ◯ Likely ◯ Very likely ◯ Certain

3. What time is 2 hours later than 11:00 pm?

   ◯ 13 o'clock ◯ 1:00 am ◯ 1:00 pm ◯ 2:11 pm

SCORE /3 (0-1) (2) (3)

TARGETING MATHS HOMEWORK: YEAR 4 © PASCAL PRESS ISBN: 9781925726572

AC9M4M03

A Visit to Fantasy World

Trains to and from Fantasy World leave every hour on the hour and the journey takes exactly 3 hours.

Jo arrived at Fantasy World at 3:00 pm and didn't get back to town until 1:00 am.

Zoe caught the 10:00 am train and spent 5 hours at Fantasy World.

Kai was halfway to Fantasy World half an hour past midday and caught the 8:00 pm train back to town.

Show on the timelines and in the chart below when these students were at Fantasy World and which trains they caught to get there and back.

1. Jo's timeline

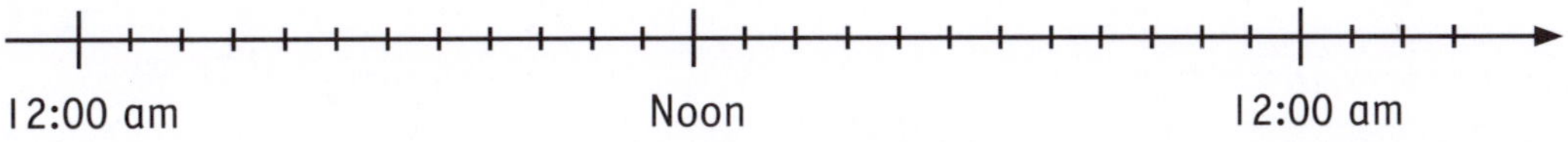

2. Zoe's timeline

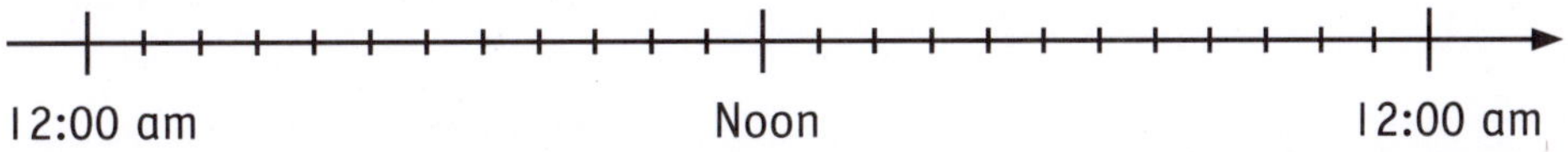

3. Kai's timeline

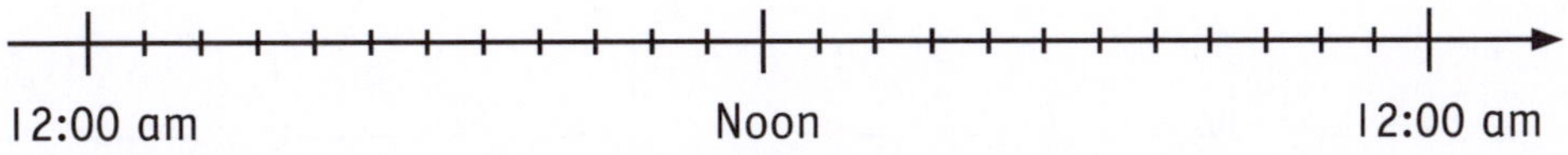

4. Complete the chart.

	Time left town	At Fantasy World			Time arrived back in town
		Arrive	Time spent	Leave	
Jo					
Zoe					
Kai					

5. Jo said she had the most time at Fantasy World. Was that true? _____

AC9M4N08

Multiplying Money Amounts

How many?

1. 50c coins in $1 ___
2. 20c coins in $1 ___
3. 10c coins in $1 ___
4. 5c coins in $1 ___

If I had to work out the cost of 4 apples at 55c each, I would split the amount 55c into 50c and 5c and then multiply by 4. Like this:

× 4 ↓ ↓ × 4

$2 + 20c = **$2·20**

Use amount-splitting for these multiplications.

5. 6 at 60c each
   ☐ + ☐
   × 6 ↓ ↓
   ☐ + ☐ = ☐

6. 3 at 70c each
   ☐ + ☐
   × 3 ↓ ↓
   ☐ + ☐ = ☐

7. 5 at 65c each
   ☐ + ☐ + ☐
   × 5 ↓ ↓ ↓
   ☐ + ☐ + ☐ = ☐

8. 4 at 75c each
   ☐ + ☐ + ☐
   × 4 ↓ ↓ ↓
   ☐ + ☐ + ☐ = ☐

9. 4 at 45c each
   ☐ + ☐ + ☐
   × 4 ↓ ↓ ↓
   ☐ + ☐ + ☐ = ☐

10. 8 at 35c each
    ☐ + ☐ + ☐
    × 8 ↓ ↓ ↓
    ☐ + ☐ + ☐ = ☐

SCORE /10 0-4 5-8 9-10

Quick Check

1. 6 at 50c each = ☐
2. 7 at 20c each = ☐
3. 6 at 15c each = ☐
4. 10 at 10c each = ☐
5. 10 at 5c each = ☐
6. 6 at 20c each = ☐
7. 3 at 50c each = ☐
8. 4 at 30c each = ☐
9. 7 at 20c each = ☐
10. 10 at 20c each = ☐

SCORE /10 0-4 5-8 9-10

TARGETING MATHS HOMEWORK: YEAR 4 © PASCAL PRESS ISBN: 9781925726572

Number & Algebra

AC9M4A02

Multiplication Facts

If you know your 5 times table, you can use it to build the 6 times table and 7 times table. This tally diagram shows how.

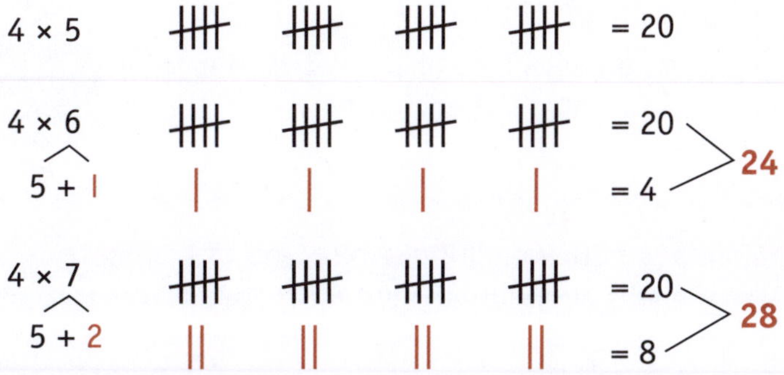

Use the 5 times table to help with these multiplication facts.

1. 6 × 6 = 6 × 5 + 6 × 1 = 30 + ☐ = ☐
2. 6 × 7 = 6 × 5 + ☐ × ☐ = ☐ + ☐ = ☐
3. 8 × 6 = 8 × 5 + ☐ × ☐ = ☐ + ☐ = ☐
4. 9 × 7 = ☐ × ☐ + ☐ × ☐ = ☐ + ☐ = ☐
5. 7 × 8 = ☐ × ☐ + ☐ × ☐ = ☐ + ☐ = ☐

SCORE /5 0-2 3-4 5

Statistics & Probability

AC9M4ST01

Midday and Midnight Temperatures

Use the graph to answer the questions.

1. What was the temperature on the coldest night?

   ____

2. What was the temperature on the coldest day?

   ____

3. What was the temperature on Thursday at midnight?

   ____

4. On which day was the temperature difference between night and day the greatest?

   ____

5. On which day was the temperature difference between night and day the least?

   ____

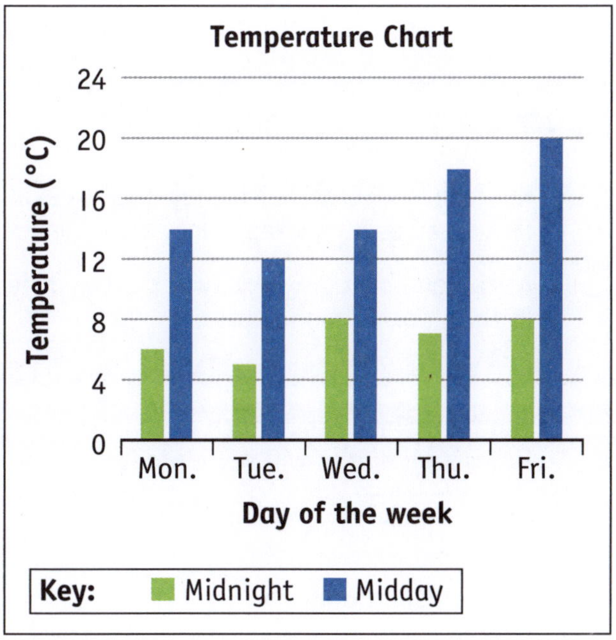

SCORE /5 0-2 3-4 5

AC9M4SP03

Lines of Symmetry

This beetle has one line of symmetry. The parts on each side of the dotted line are a mirror image of each other.

The shapes below each have one line of symmetry. Draw the line of symmetry on each shape.

1. 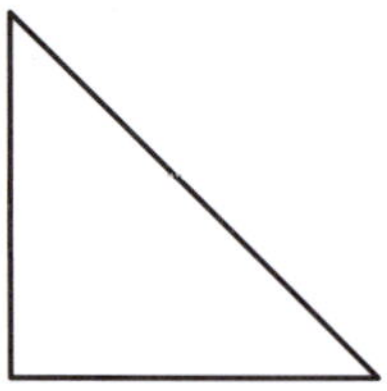
2. 
3. 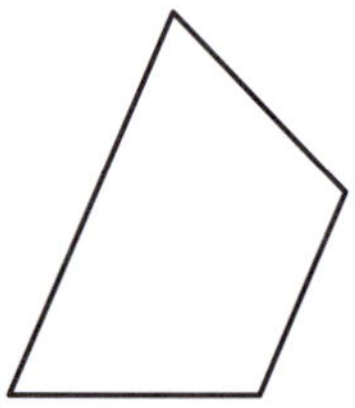
4. 
5. 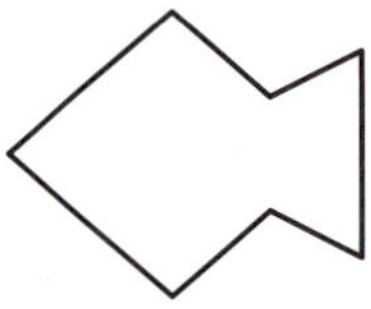
6. 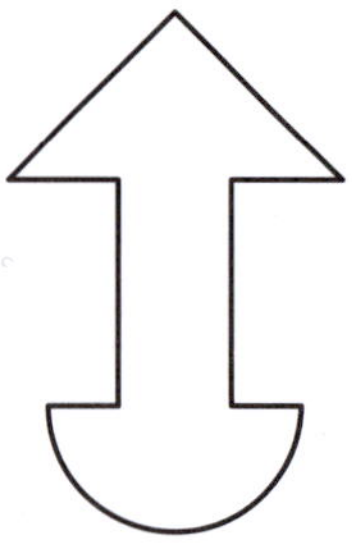
7. 
8. 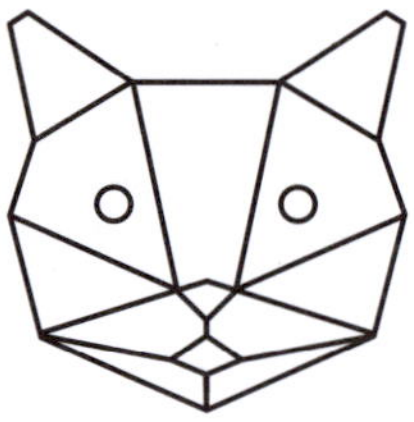
9. This shape has two lines of symmetry. Draw both lines on the shape.

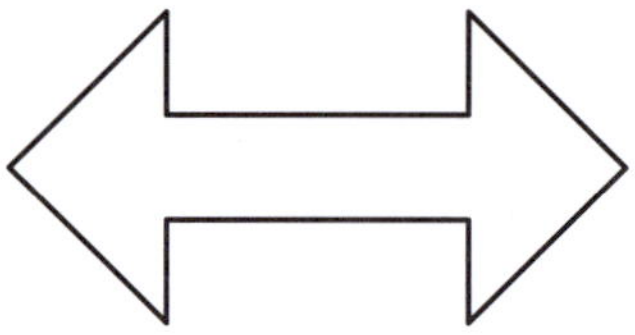

SCORE /9 0-4 5-7 8-9

NAPLAN*-style Questions

1. Kai receives $6 pocket money each week. Dan receives $8 pocket money each week.

   At the end of four weeks, what is the difference in their pocket money totals?

   ⬭ $6 ⬭ $8 ⬭ $14 ⬭ $16

2. There are 6 apples in each net. How many apples will be in 6 nets?

   ____ apples

3. Look at the graph. How many more people walked than cycled to the shop?

   [ ]

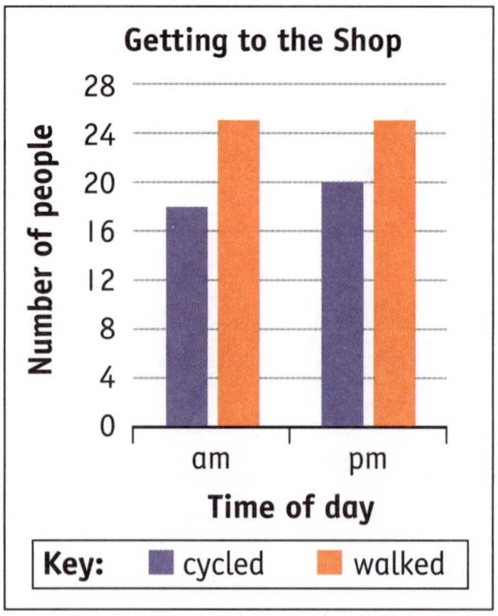

SCORE /3 0-1 2 3

TARGETING MATHS HOMEWORK: YEAR 4 © PASCAL PRESS ISBN: 9781925726572

Problem of the Week

With Seven Squares

On her grid paper, Jo has been drawing the outline of seven squares that make a shape. Then she looks to see if the shape she has drawn has a line of symmetry.

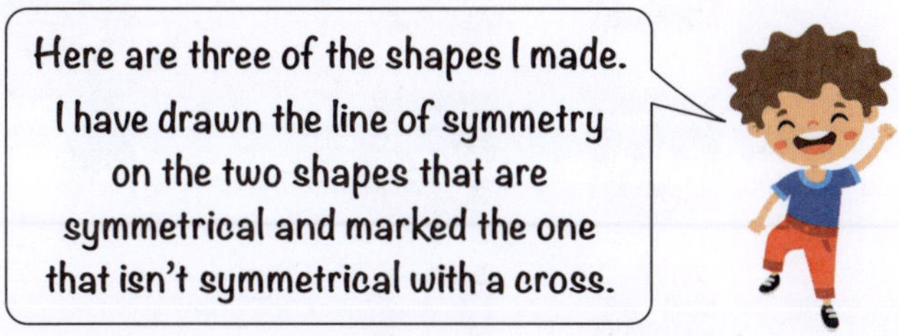

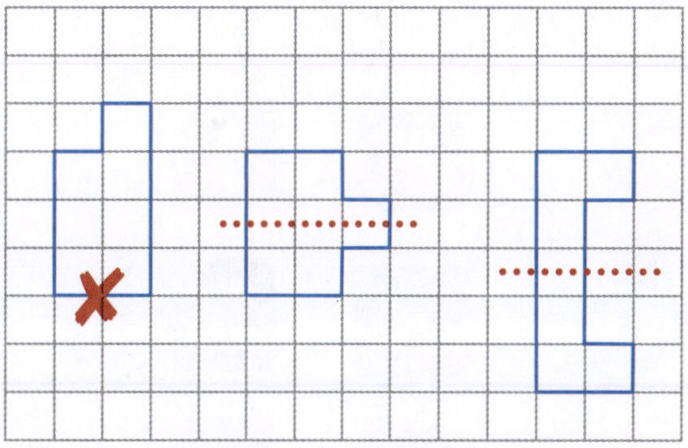

Your turn now!

1. Draw the outlines of shapes made from seven squares.
2. Show lines of symmetry or mark shapes with a cross.

3. Which type of shape did you find it easier to make?

◯ A shape with symmetry

◯ A shape with no symmetry

UNIT 7

Number & Algebra

AC9M4N02

Odd and Even Numbers

Zoe and Kai were rolling two dice to make each score for their game of Snakes and Ladders.

Zoe made these scores.

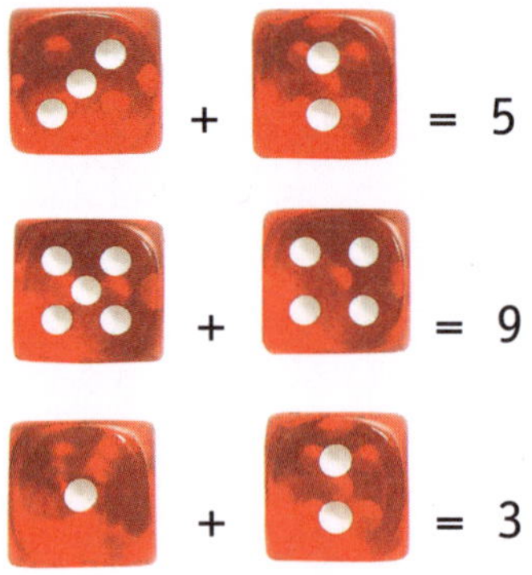

Kai made these scores.

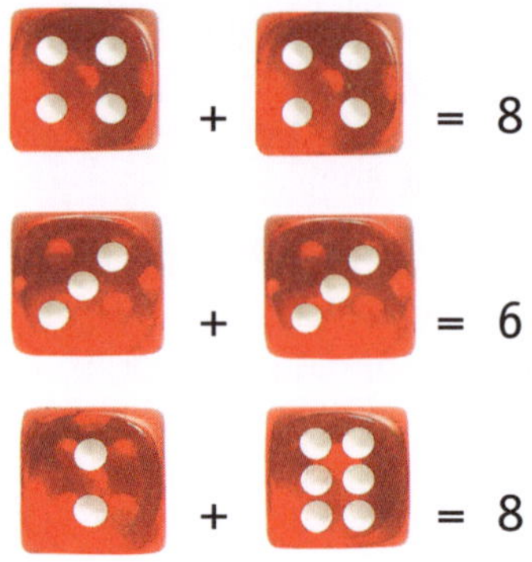

① What is special about the two numbers that Zoe rolled each time and the totals that she made?

____

② What is special about the two numbers that Kai rolled each time and the totals that he made?

____

③ Show three rolls and dice totals that could be made with all odd number dice rolls.

☐ + ☐ = ____ ☐ + ☐ = ____ ☐ + ☐ = ____

④ Show three rolls and dice totals that could be made with all even number dice rolls.

☐ + ☐ = ____ ☐ + ☐ = ____ ☐ + ☐ = ____

⑤ Show three rolls and dice totals that could be made with one even number and one odd number dice roll.

☐ + ☐ = ____ ☐ + ☐ = ____ ☐ + ☐ = ____

SCORE /5 0-2 3-4 5

Quick Check

Write O for odd and E for even in the box to describe the totals of these addition pairs.
You don't need to work out the additions.
Just use the odd and even rule to describe the totals.

① 6 + 6 ☐

② 5 + 7 ☐

③ 7 + 7 ☐

④ 3 + 5 ☐

⑤ 13 + 13 ☐

⑥ 9 + 9 ☐

⑦ 27 + 35 ☐

⑧ 46 + 30 ☐

⑨ 25 + 25 ☐

⑩ 18 + 23 ☐

SCORE /10 0-4 5-8 9-10

TARGETING MATHS HOMEWORK: YEAR 4 © PASCAL PRESS ISBN: 9781925726572

Number & Algebra

AC9M4N02, AC9M4N09

Number Sequences

Write the next three numbers in these patterns.
Then write the rule and say whether the 15th number will be odd or even.
The first one has been started for you.

	Sequence	Rule	Odd or Even?
1	1, 4, 3, 6, 5, ___, ___, ___, ...	+ 3, − 1	Odd
2	2, 4, 3, 5, 4, ___, ___, ___, ...		
3	3, 5, 4, 6, 5, ___, ___, ___, ...		
4	2, 8, 14, 20, 26, ___, ___, ___, ...		
5	3, 8, 13, 18, 23, ___, ___, ___, ...		
6	4, 8, 5, 9, 6, ___, ___, ___, ...		
7	13, 12, 14, 13, 15, ___, ___, ___, ...		
8	5, 15, 13, 23, 21, ___, ___, ___, ...		
9	2, 4, 8, 16, 32, ___, ___, ___, ...		
10	32, 41, 39, 48, 46, ___, ___, ___, ...		

SCORE /10 0-4 5-8 9-10

Statistics & Probability

AC9M4ST02

Yesterday's Weather

Line Graph

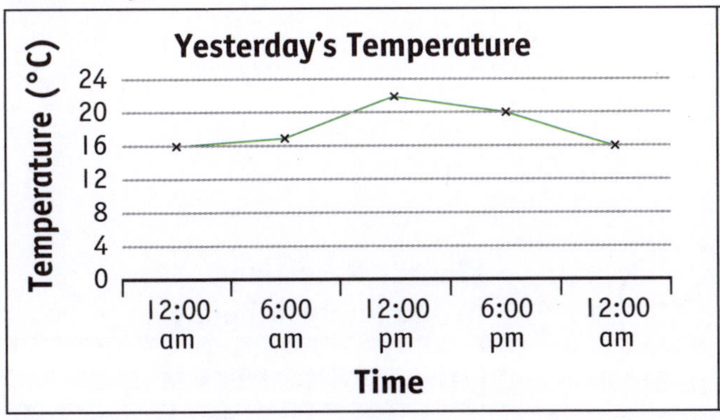

Column Graph

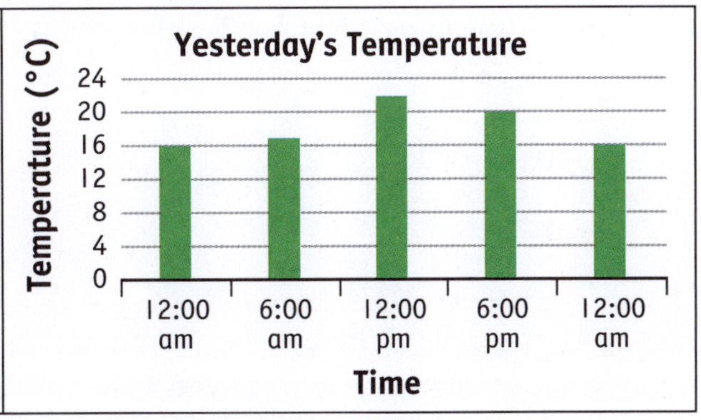

1. What was the temperature at midnight yesterday? ________
2. During which timespan did it warm the most? Between ______ and ______
3. During which timespan did it cool down the least? Between ______ and ______
4. Which graph do you think shows the temperature changes in the better way? ______________
5. What makes you think that? ______________________________

SCORE /5 0-2 3-4 5

Measurement & Space

AC9M4M04

All About Angles

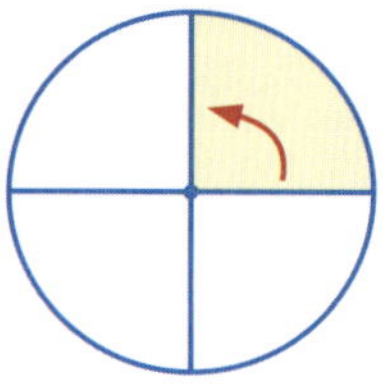

A right angle is like a $\frac{1}{4}$ turn.

A straight angle is like a $\frac{1}{2}$ turn.

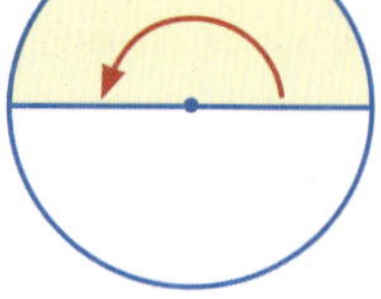

An acute angle is smaller than a right angle.
An obtuse angle is larger than a right angle.
A reflex angle is larger than a straight angle.

Name each of these angles.

1.  ____________ angle
2. 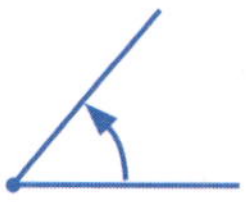 ____________ angle
3.  ____________ angle
4.  ____________ angle
5. 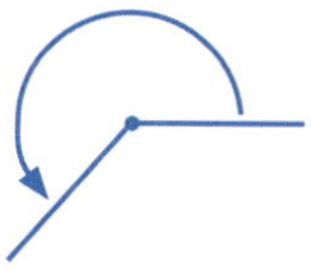 ____________ angle

Write the letter of each angle next to its name:

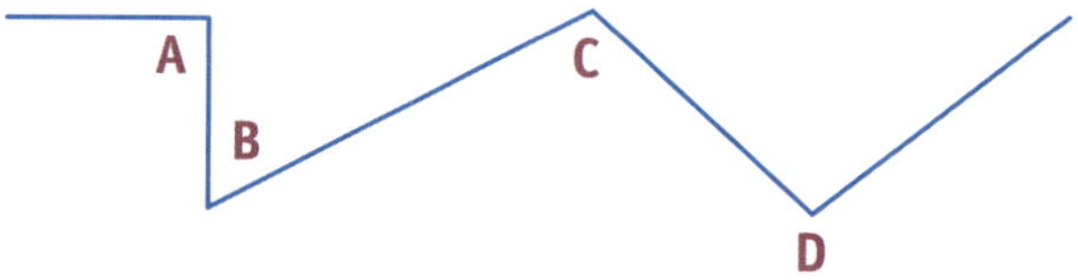

6. ☐ obtuse
7. ☐ right
8. ☐ acute
9. ☐ reflex

SCORE /9 0-4 5-7 8-9

NAPLAN*-style Questions

1. What is the largest odd number that can be made with these digits? **3 4 5**

☐

2. Which diagram shows an acute angle?

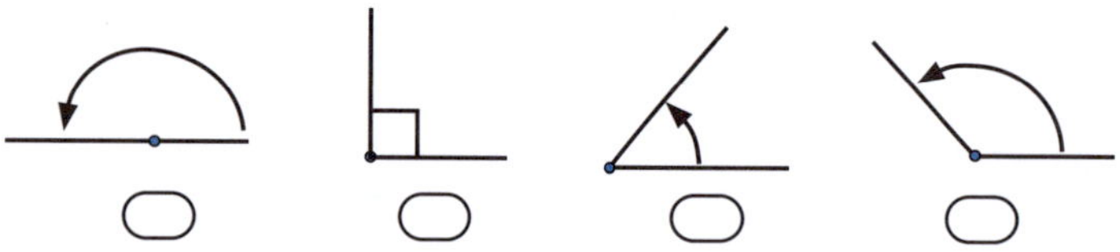

3. Look at the chart. How much plastic altogether was processed in Week 1?

☐ tonnes

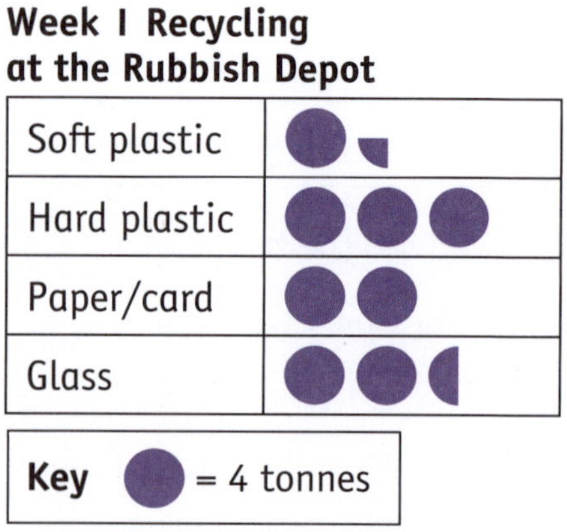

SCORE /3 0-1 2 3

TARGETING MATHS HOMEWORK: YEAR 4 © PASCAL PRESS ISBN: 9781925726572

Problem of the Week

AC9M4M04

Crossing Lines

Each diagram has been drawn with three straight lines. Each line crosses at least one other line.

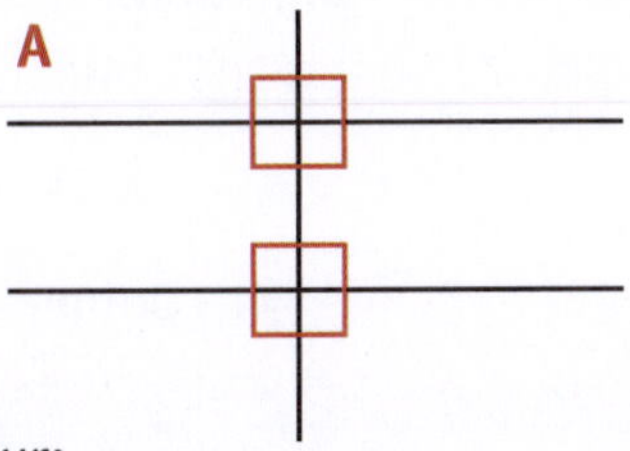

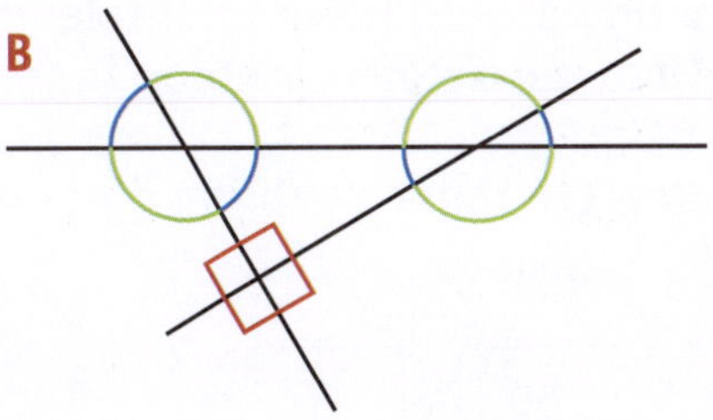

The angles have been marked in colour:

- right angle
- acute angle
- obtuse angle

The chart below shows the properties of the angles in each shape.

	Total number of angles	Right angles	Acute angles	Obtuse angles
Shape A	8	8	0	0
Shape B	12	4	4	4

Draw some diagrams with three crossing lines of your own.
Write your angle information in the table below.

	Total number of angles	Right angles	Acute angles	Obtuse angles

Use your information to answer these questions.

1. Will there always be 8 or 12 angles? ______
2. Will there always be some right angles? ______
3. Will there always be 4 or 8 angles that are right, acute or obtuse? ______

AC9M4N06

Halves and Doubles

Being able to halve or double a number can be really helpful when multiplying.
Here we explore a method for doubling and halving that makes the process clear.

Complete these doubling diagrams.

① Double 63.

60 + 3
↓ Double ↓
120 + 6 = ____

② Double 76.

____ + ____
↓ ________ ↓
____ + ____ = ____

③ Double 39.

____ + ____
↓ ________ ↓
____ + ____ = ____

④ Double 163.

160 + ____
↓ ________ ↓
____ + ____ = ____

Complete these finding halves diagrams.

⑤ Halve 66.

60 + 6
↓ Halve ↓
30 + 3 = ____

⑥ Halve 84.

____ + ____
↓ ________ ↓
____ + ____ = ____

⑦ Halve 94.

____ + ____
↓ ________ ↓
____ + ____ = ____

⑧ Halve 168.

____ + ____
↓ ________ ↓
____ + ____ = ____

SCORE /8 0-3 4-6 7-8

Quick Check

You can draw pictures to help.

Find doubles.

① 28 ______
② 36 ______
③ 79 ______
④ 58 ______
⑤ 123 ______

Find halves.

⑥ 24 ______
⑦ 30 ______
⑧ 36 ______
⑨ 50 ______
⑩ 70 ______
⑪ 90 ______
⑫ 74 ______
⑬ 58 ______
⑭ 112 ______
⑮ 124 ______

SCORE /15 0-7 8-11 12-15

TARGETING MATHS HOMEWORK: YEAR 4 © PASCAL PRESS ISBN: 9781925726572

Number & Algebra

AC9M4N06

Multiplication

In this example, you can see that multiplying by 5 gives half the answer of multiplying by 10.

10 × 4 = 40
5 × 4 = 20

For 5 × 84,
do 10 × 84 = 840
and halve the answer:
5 × 84 = 420

Try using Chen's strategy with these examples.

① For 5 × 26,
do 10 × 26 = 260 and halve:
5 × 26 = 130

② For 5 × 44,
do ☐ × ☐ = ☐ and halve:
5 × ☐ = ☐

③ For 5 × 38,
do ☐ × ☐ = ☐ and halve:
☐ × ☐ = ☐

④ For 5 × 62,
do ☐ × ☐ = ☐ and halve:
☐ × ☐ = ☐

⑤ For 5 × 96,
do ☐ × ☐ = ☐ and halve:
☐ × ☐ = ☐

⑥ For 5 × 64,
do ☐ × ☐ = ☐ and halve:
☐ × ☐ = ☐

⑦ For 5 × 76,
do ☐ × ☐ = ☐ and halve:
☐ × ☐ = ☐

⑧ For 5 × 54,
do ☐ × ☐ = ☐ and halve:
☐ × ☐ = ☐

SCORE /8 0-3 4-6 7-8

Measurement & Space

AC9M4SP01

Scale it Up

Make the robot taller.
The head has been started for you.

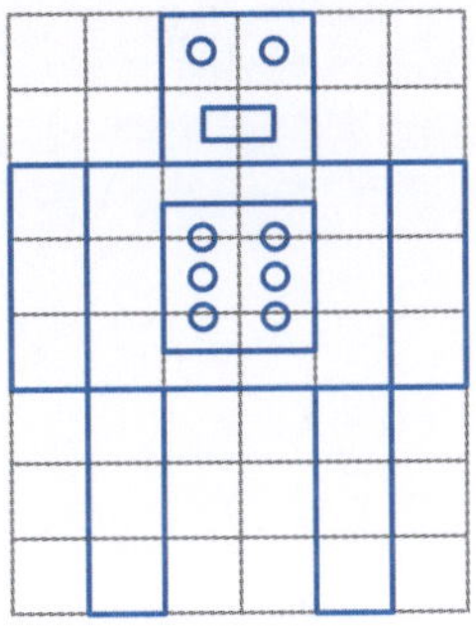

AC9M4P02

North, East, South or West?

I have been gently dropping a pencil, point down, onto the red circle and recording the direction in which it fell.

I was sure that there would be the same chance of the pencil dropping in each direction.

The chart shows the results of my first 8 trials.

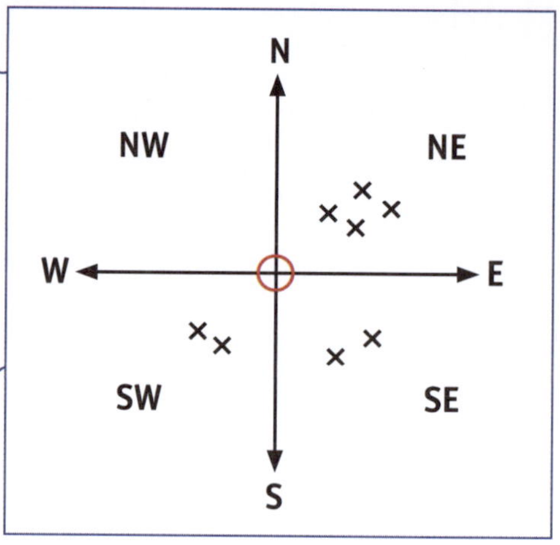

① Was Zoe correct in her thinking? ______

② Zoe drops the pencil 8 more times.
Do you think the chance of a pencil landing in the NW quadrant is **unlikely**? ______

Why? ______________________________

③ What do you think the results of her next 8 drops will be?

NE ☐ SE ☐ SW ☐ NW ☐

Make yourself a NESW diagram with a circle in the middle.
Drop a pencil 8 times, making sure that it lands on your red circle in the middle.

④ Record your results here.

NE ☐ SE ☐ SW ☐ NW ☐

⑤ Were your results the same as Zoe's? ______

Why was that? ______________________________

SCORE /5 0-2 3-4 5

NAPLAN*-style Questions

① Colour the arrow that points to the number that is halfway between 15 and 35.

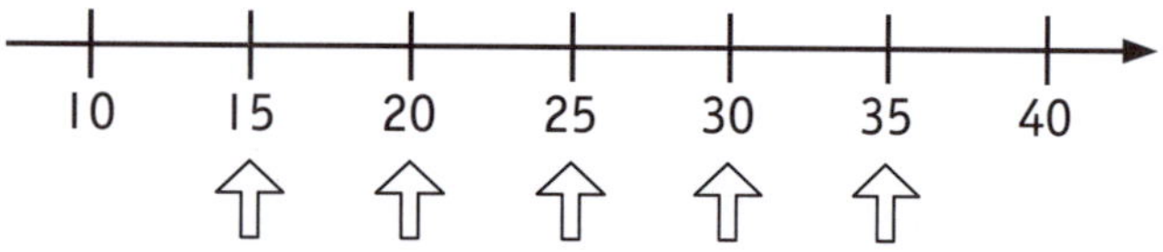

② Five shirts at $35 each will cost how many dollars?

$______

③ What is the chance that this spinner will land on blue?

- ◯ Impossible
- ◯ Not likely
- ◯ Likely
- ◯ Very likely
- ◯ Certain

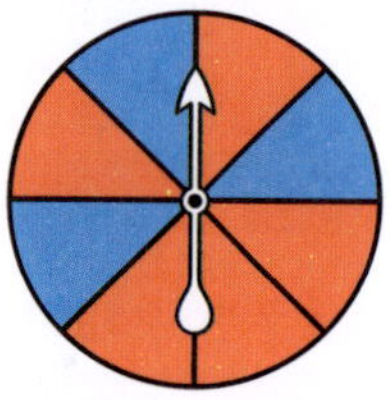

SCORE /3 0-1 2 3

TARGETING MATHS HOMEWORK: YEAR 4 © PASCAL PRESS ISBN: 9781925726572

Problem of the Week

AC9M4P01

Heads or Tails

Claire, Jake and Kai each have a 20c coin to toss.
Their job is to say what they think is the most likely outcome.

Claire, Jake and Kai each have a prediction.

Claire:
I think it is impossible to get three heads or three tails.

Jake:
I think it is most likely to get two heads and one tail.

Kai:
I think it is most likely to get one head and two tails.

1. Make a list of all the possible outcomes of tossing three coins.
Here is one to start you off!

2. Toss three coins 10 times and record your results in this table.

	All Heads	**All Tails**	**Two Heads, One Tail**	**One Head, Two Tails**
Test				

3. If you were to repeat the 10-times experiment, what do you think your results would be?

	All Heads	**All Tails**	**Two Heads, One Tail**	**One Head, Two Tails**
Think				

4. Whose prediction do you think was the best?
◯ Claire ◯ Jake ◯ Kai

TARGETING MATHS HOMEWORK: YEAR 4 © PASCAL PRESS ISBN: 9781925726572

Review: Unit 1 and Unit 2

AC9M4N01, AC9M4N02, AC9M4N03, AC9M4N04, AC9M4N05, AC9M4M04, AC9M4P01

Write these fractions as decimals.

(1) $\frac{6}{10}$ = __·__

(2) $\frac{3}{10}$ = __·__

(3) $\frac{12}{10}$ = __·__

Write these decimals as fractions.

(4) 0·7 = —

(5) 0·4 = —

(6) 1·6 = —

(7) Circle the even numbers.

3 21 12 32 42 50

(8) Which two shapes can be made with one rectangle and one triangle?

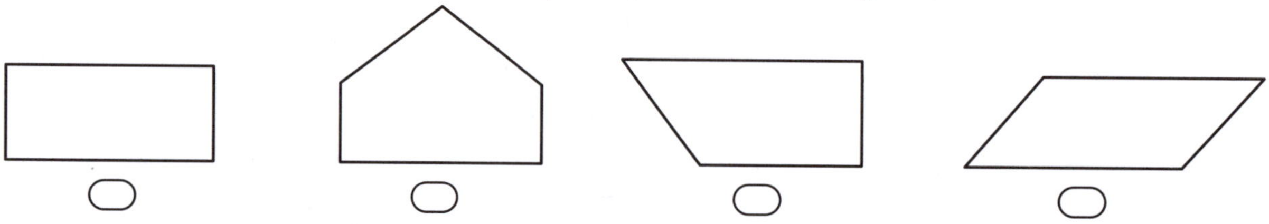

Write the missing fractions.

(9)

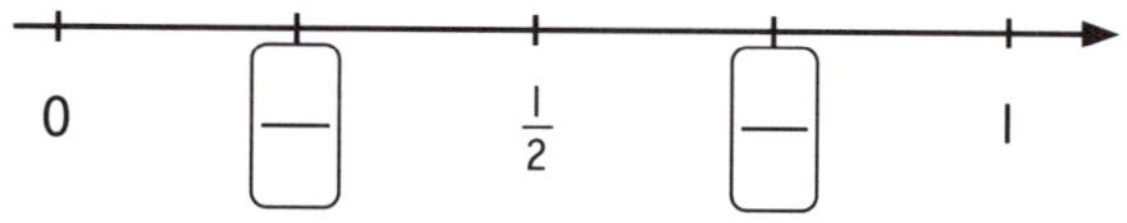

(10)

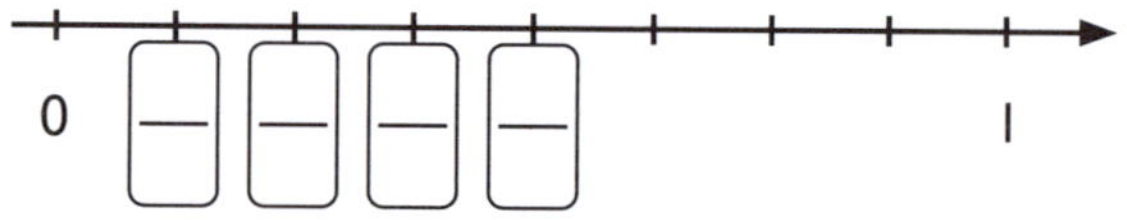

(11) Complete the equivalent fraction: $\frac{3}{4} = \frac{}{8}$

What will these numbers be after multiplying by 10?

(12) 32 ⇨ ________

(13) 9 ⇨ ________

(14) 70 ⇨ ________

(15) 125 ⇨ ________

(16) 190 ⇨ ________

(17) 400 ⇨ ________

What will these numbers be after multiplying by 100?

(18) 5 ⇨ ________

(19) 75 ⇨ ________

(20) 40 ⇨ ________

(21) 820 ⇨ ________

(22) Which spinner has the better chance of spinning red?

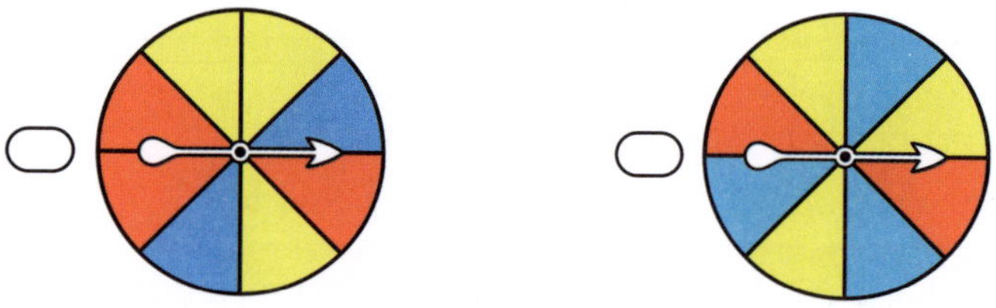

SCORE /22 0-10 11-20 21-22

TARGETING MATHS HOMEWORK: YEAR 4 © PASCAL PRESS ISBN: 9781925726572

Review: Unit 3 and Unit 4

AC9M4N04, AC9M4N05, AC9M4N06, AC9M4ST03, AC9M4SP02

Round to the nearest 10.

1. 52 ______ 2. 35 ______ 3. 49 ______ 4. 98 ______

Round to the nearest 100.

5. 134 ______ 6. 298 ______ 7. 345 ______ 8. 988 ______

Complete the bar models for these questions.

9. There were 6 apples in the bowl this morning but there are only 2 left now. How many did Zoe eat?

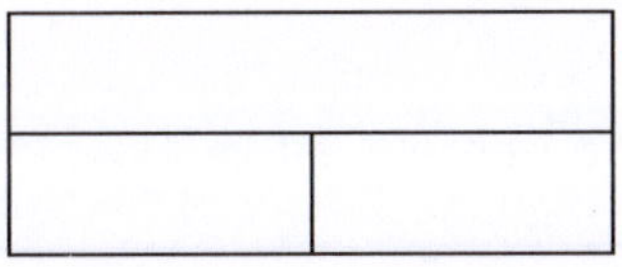

10. Dan used 8 apples to make an apple pie. Now there are only 2 left in the bag. How many apples were in the bag that Dan bought?

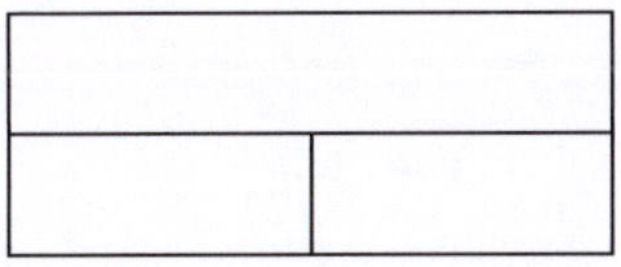

Jo made these shapes with 1 cm squares. What is the perimeter?

11. P = ________

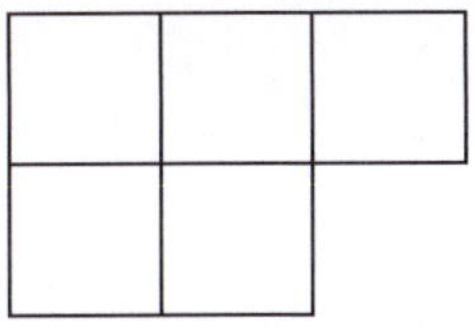

12. P = ________

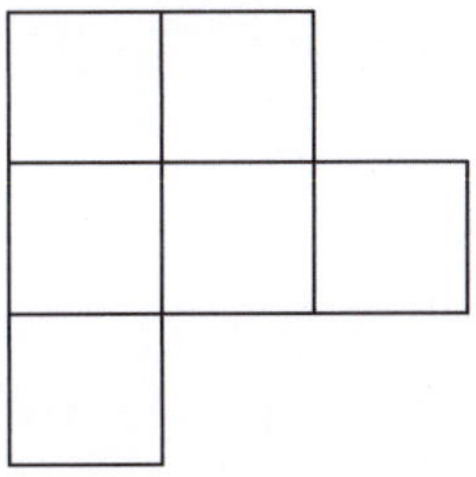

13. P = ________

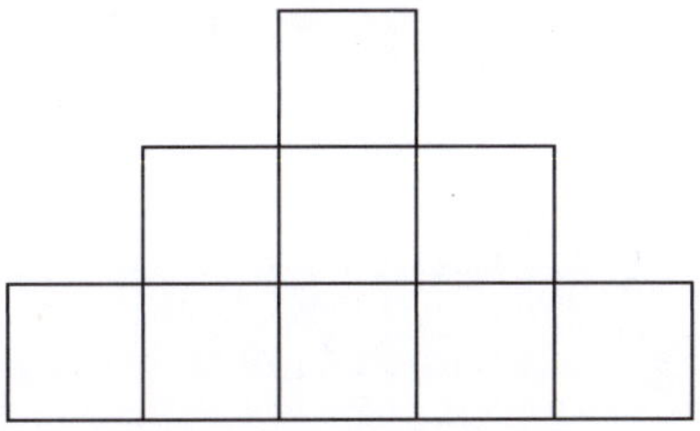

14. This is a chart of a survey of 35 people to find out favourite cordial flavours. Complete the key.

Lemon	
Mango	
Orange	
Lime	

Key  = ___ votes

Show two different ways of rounding and estimating. Then check on your calculator.

15. 243 + 643
Estimate 1 ______________________
Estimate 2 ______________________
Check ______________________

16. 355 + 424
Estimate 1 ______________________
Estimate 2 ______________________
Check ______________________

17. 372 + 476
Estimate 1 ______________________
Estimate 2 ______________________
Check ______________________

18. 465 + 726
Estimate 1 ______________________
Estimate 2 ______________________
Check ______________________

19. After spending $73, Jake had $196 left. How much money did he start with? Write the calculator sequence and the answer.

___ [+] ___ [=] ___

SCORE | /19 | 0-9 | 10-17 | 18-19

Review: Unit 5 and Unit 6

AC9M4N04, AC9M4N07, AC9M4N08, AC9M4A02, AC9M4M03, AC9M4SP03

Use different colours to show these additions on the grids.

1. $\frac{4}{10} + 0{\cdot}5 + \frac{5}{10}$

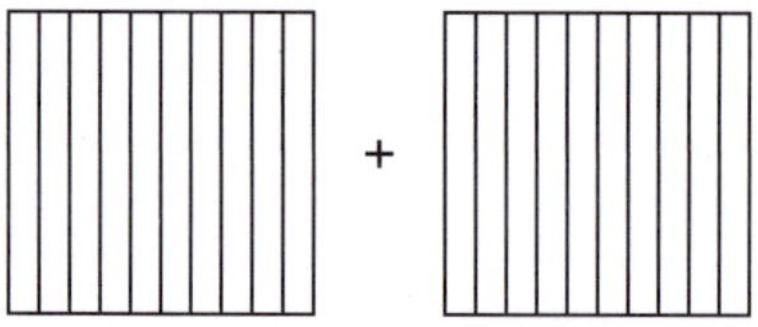

2. $\frac{6}{10} + \frac{6}{10} + 0{\cdot}3$

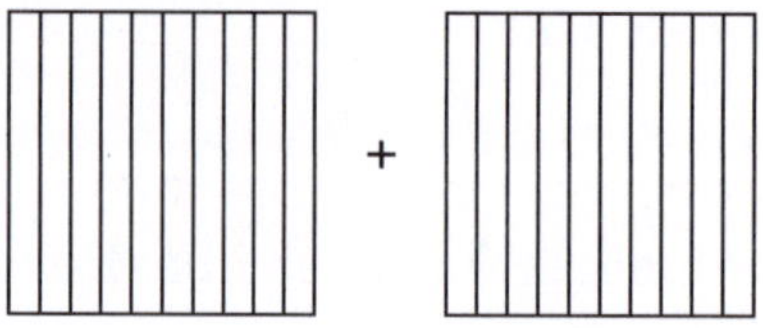

Use rounding to the nearest 10 to estimate these subtractions and then check with a calculator.

3. **Find** 546 – 327

Estimate _____ – _____ = _____

Check Actual answer = _____

4. **Find** 1373 – 899

Estimate _____ – _____ = _____

Check Actual answer = _____

Mark these times on the timeline.

5. **A** 9:00 pm

B 10:00 am

C midday

Midnight 12 o'clock — Midnight 12 o'clock

Use amount-splitting for these multiplications.

6. 6 at 25c each

☐ + ☐

× __

☐ + ☐ = ☐

7. 4 at 65c each

☐ + ☐ + ☐

× __

☐ + ☐ + ☐ = ☐

Build on the 5 times table for these calculations.

8. 6 × 8 = 6 × 5 + 6 × ☐

= ☐ + ☐ = ☐

9. 7 × 6 = ☐ × ☐ + ☐ × ☐

= ☐ + ☐ = ☐

10. 8 × 9 = ☐ × ☐ + ☐ × ☐

= ☐ + ☐ = ☐

Draw the lines of symmetry on these shapes.

11.

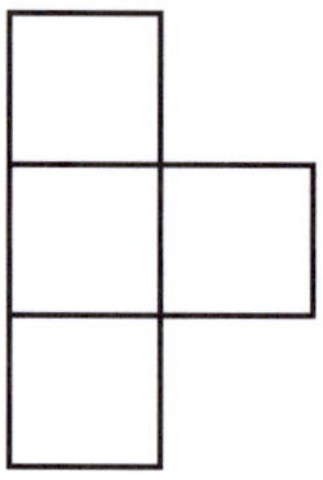

12.

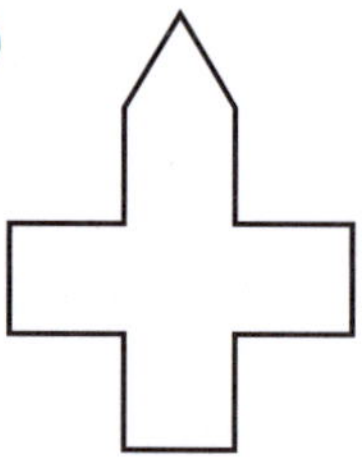

13.

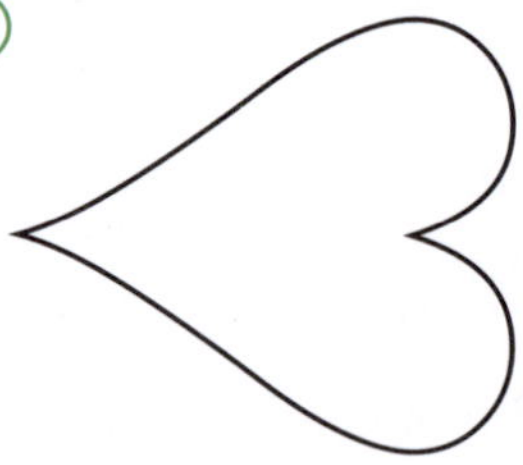

SCORE /13 0-6 7-11 12-13

TARGETING MATHS HOMEWORK: YEAR 4 © PASCAL PRESS ISBN: 9781925726572

Review: Unit 7 and Unit 8

AC9M4N02, AC9M4N04, AC9M4N06, AC9M4N09, AC9M4M04, AC9M4ST02

Write **odd** or **even** for these totals.

1. 6 + 7 is _______
2. 13 + 24 is _______
3. 45 + 43 is _______
4. 27 + 85 is _______
5. 126 + 240 is _______
6. 1082 + 1034 is _______

Continue these patterns and write the pattern rule.

7. 4, 7, 5, 8, ___, ___, ___ The rule is _______
8. 12, 17, 22, 27, ___, ___, ___ The rule is _______
9. 58, 55, 56, 53, 54, ___, ___, ___ The rule is _______

Use the graph to answer these questions.

10. How much hotter was it at noon yesterday than at the previous midnight?

    ___ degrees Celsius

11. What was the coolest temperature?

    ___ degrees Celsius

12. How much hotter was it at 6:00 pm than at 6:00 am?

    ___ degrees Celsius

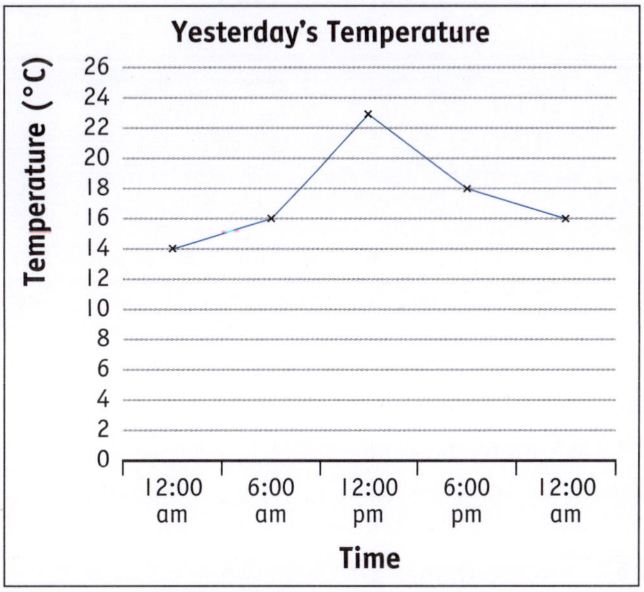

Name the angles.

13. _____________ angle

14. _____________ angle

15. _____________ angle

Double these numbers.

16. Double 36 is _____.
17. Double 49 is _____.
18. Double 109 is _____.

Halve these numbers.

19. Half of 28 is _____.
20. Half of 54 is _____.
21. Half of 208 is _____.

Complete these multiplications.

22. For 5 × 32,

    do ☐ × ☐ = ☐ and halve:

    5 × ☐ = ☐

23. For 5 × 46,

    do ☐ × ☐ = ☐ and halve:

    ☐ × ☐ = ☐

SCORE /23 0-11 12-21 22-23

AC9M4N01, AC9M4N03

Counting in Tenths

Counting in tenths helps with counting in decimals.

$\frac{1}{10}$	$\frac{2}{10}$	$\frac{3}{10}$	$\frac{4}{10}$	$\frac{5}{10}$
0·1	0·2	0·3	0·4	0·5

Continue these sequences. Remember that $\frac{10}{10}$ is equal to 1.

1	$\frac{7}{10}$	$\frac{8}{10}$	__	1	$1\frac{1}{10}$	__	__	__	__
	0·7	0·8	0·9	__·__	__·__	__·__	__·__	__·__	__·__
2	$\frac{15}{10}$	$\frac{16}{10}$	__	__	__	__	__	__	__
	1·5	__·__	__·__	__·__	__·__	__·__	__·__	__·__	__·__
3	$\frac{5}{10}$	$\frac{7}{10}$	$\frac{9}{10}$	__	__	__	__	__	__
	0·5	0·7	__·__	__·__	__·__	__·__	__·__	__·__	__·__

Count in tenths to answer these questions.
Write your answers as decimals.

4. 0·3 more than 0·6 ____________
5. $\frac{5}{10}$ more than 0·6 ____________
6. $\frac{5}{10}$ more than 1·3 ____________
7. $\frac{3}{10}$ less than 1·1 ____________
8. 0·6 less than 1·1 ____________
9. 0·3 less than 2·2 ____________
10. $\frac{13}{10}$ less than 2 ____________

SCORE /10 0-4 5-8 9-10

Quick Check

Write the decimal that is $\frac{3}{10}$ more.

1. $\frac{5}{10}$ __·__
2. 1 __·__
3. 0·4 __·__
4. 0·8 __·__
5. 1·9 __·__

Write the decimal that is $\frac{3}{10}$ less.

6. $\frac{9}{10}$ __·__
7. 1·2 __·__
8. 0·8 __·__
9. 2·1 __·__
10. $\frac{15}{10}$ __·__

SCORE /10 0-4 5-8 9-10

TARGETING MATHS HOMEWORK: YEAR 4 © PASCAL PRESS ISBN: 9781925726572

Number & Algebra

AC9M4N04

All About Fractions

Fractions are made with two parts, the **numerator** and the **denominator**.

For the fraction $\frac{3}{8}$, the number 3 is above the **fraction bar** (or vinculum) and 3 is the numerator.

The number 8 is below the fraction bar and is the denominator.

When the numerator is smaller than the denominator, as in $\frac{3}{8}$, the fraction is called a **proper fraction**.

If the numerator is larger than the denominator, as in $\frac{8}{3}$, the fraction is called an **improper fraction**.

$\frac{3}{8}$	3 ← Numerator; — ← Fraction Bar; 8 ← Denominator
$\frac{3}{8}$	Proper Fraction
$\frac{8}{3}$	Improper Fraction
$2\frac{2}{3}$	Mixed Number

Continue these fraction counting sequences using only proper fractions or mixed numbers.

1. $\frac{2}{4}$, $\frac{3}{4}$, 1, $1\frac{1}{4}$, 1—, 1—, 2, 2—, 2—, 2—, __
2. $\frac{1}{3}$, $\frac{2}{3}$, 1, 1—, 1—, 2, 2—, 2—, __, __—, __—, __
3. $\frac{3}{5}$, $\frac{4}{5}$, 1, __—, __—, __—, __—, __, __—, __—

Continue these fraction counting sequences using only proper or improper fractions.

4. $\frac{2}{4}$, $\frac{3}{4}$, $\frac{4}{4}$, $\frac{5}{4}$, —, —, —, —, —, —
5. $\frac{1}{3}$, $\frac{2}{3}$, —, —, —, —, —, —, —, —
6. $\frac{3}{5}$, $\frac{4}{5}$, —, —, —, —, —, —, —, —

SCORE /6 0-2 3-4 5-6

Statistics & Probability

AC9M4ST01

Fast-food Survey

Jake collected data from 100 people about a famous fast-food place.

He asked people, "Do you love it, hate it or don't really care?"

He found that four-tenths loved it, three-tenths hated it, and the rest didn't really care.

He began this graph of his data.

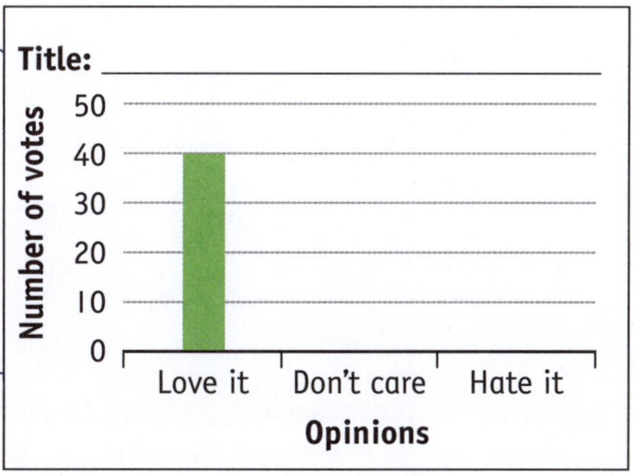

1. Complete the graph and give it a title.

Answer these questions about the graph.

2. How many people loved the fast-food place? _____
3. How many people hated the fast-food place? _____
4. How many people really didn't care? _____
5. How many more people either hated it or didn't care than loved it? _____

SCORE /5 0-2 3-4 5

Measurement & Space

AC9M4M01

Measuring in Centimetres or Millimetres

There are 10 millimetres in a centimetre.
This part of a ruler shows 2 cm (short for centimetres), which is the same as 20 mm (short for millimetres). Dressmakers mainly use centimetres when making a dress. Carpenters use millimetres when building with wood.

Write the length of each blue line in both centimetres and millimetres.

① 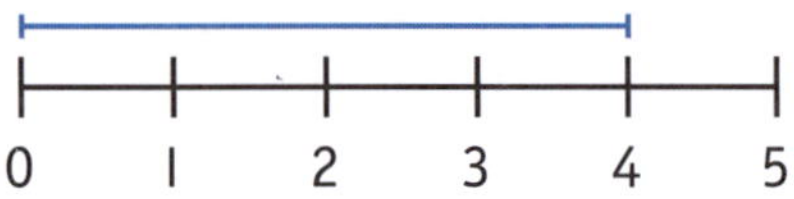

4 cm or ____ mm

② 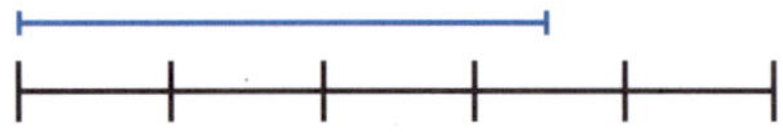

___ cm or ___ mm

③ 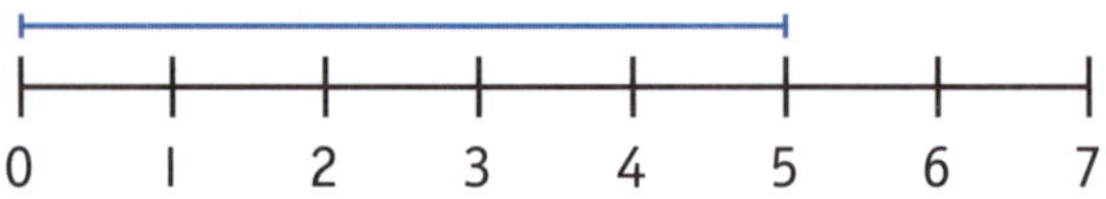

___ cm or ___ mm

④ 

___ cm or ___ mm

This one is meant to trick you. Look carefully!

⑤ 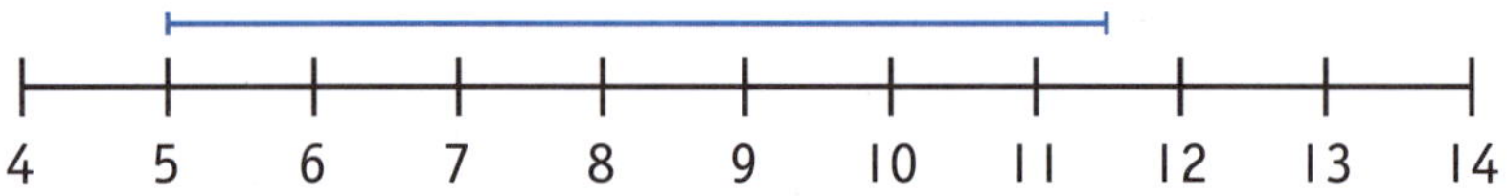

___ cm or ___ mm

Now answer these length questions.

⑥ How many centimetres in 1 metre 25 cm? ________ cm

⑦ How many millimetres in 1 metre 25 cm? __________ mm

⑧ About how long do you think this red line is? _____ cm or _____ mm

SCORE /8 0-3 4-6 7-8

NAPLAN*-style Questions

① What is the value of A?

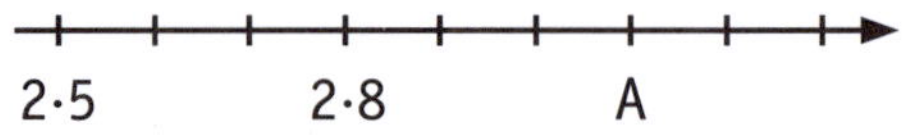

② Which of these is about 110 mm in length?

③ Look at the graph.
How much further than Jo did Kai throw the tennis ball?

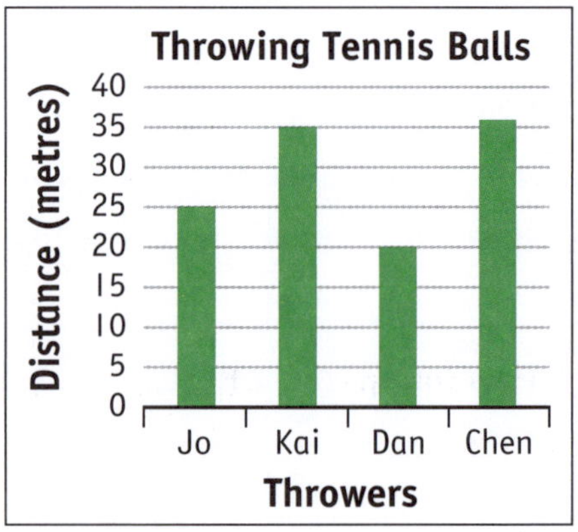

SCORE /3 0-1 2 3

TARGETING MATHS HOMEWORK: YEAR 4 © PASCAL PRESS ISBN: 9781925726572

Don't Cross the Line

Claire and her friends have made up a game they call Don't Cross the Line.

First they mark two lines, a **start line** and a **finish line**, 50 cm apart on a table.

On a turn, each player uses a 20c coin, the **pusher**, to push a 5c coin, the **slider**, from the start line towards the finish line. They are allowed four pushes.

The player whose slider is closest to the line is the winner.

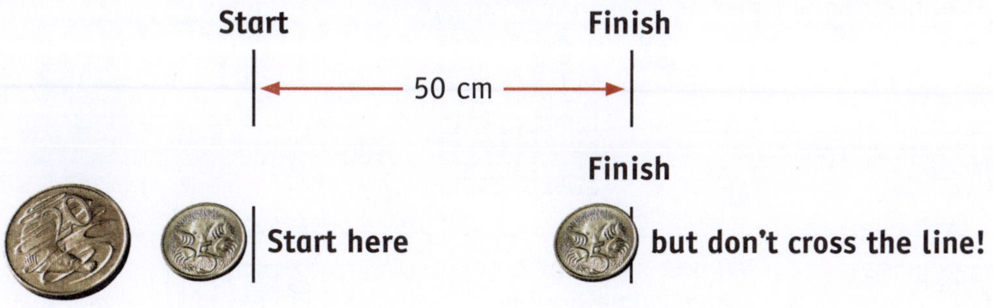

Claire's pushes were:
- Push 1: 65 mm
- Push 2: 15 cm
- Push 3: 120 mm
- Push 4: 6 cm

Jo's pushes were:
- Push 1: 135 mm
- Push 2: 125 mm
- Push 3: 7 cm
- Push 4: 13 cm

Chen's pushes were:
- Push 1: 140 mm
- Push 2: 80 mm
- Push 3: 75 mm
- Push 4: 14 cm

Zoe's pushes were:
- Push 1: 140 mm
- Push 2: 12 cm
- Push 3: 55 mm
- Push 4: 126 mm

How did they go?

1. Who had the longest single push? ____________
2. Who had the shortest single push? ____________
3. Whose total pushes put them closest to the finish line? ____________
4. Whose total pushes put them furthest from the finish line? ____________
5. How far ahead of the person in fourth place was the winner? ____________

AC9M4N01

TERM 2

Introducing Hundredths

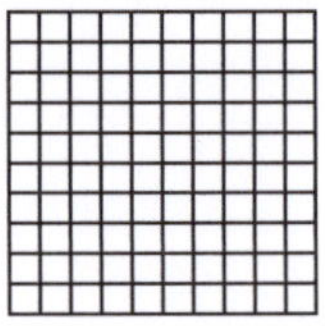

The large square has been divided into one hundred equal small squares.

Each small square is called one **hundredth**.

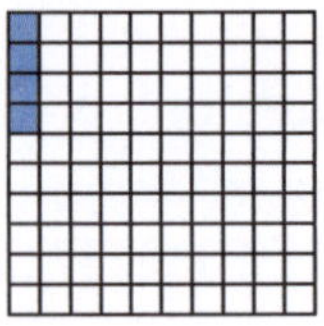

Four of the one-hundredth squares have been coloured blue to show $\frac{4}{100}$.

How many hundredths are blue in these large squares?

① 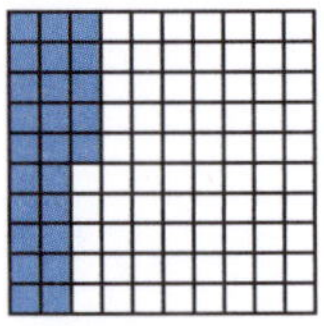 $\frac{}{100}$

② 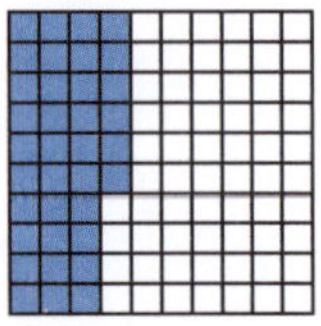 $\frac{}{100}$

③ 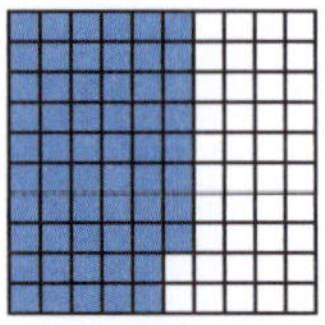 $\frac{}{100}$

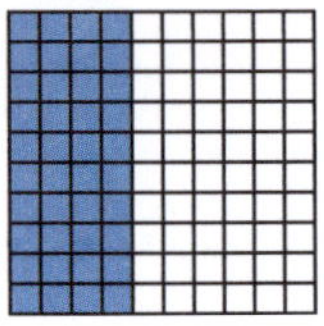

$\frac{40}{100}$

40 one-hundredths of this large square are blue.

We can see that it is the same amount of blue as $\frac{4}{10}$ of the large square.

$\frac{4}{10}$

Label these diagrams in tenths and hundredths.

④ 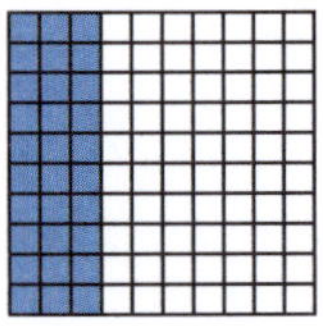 $\frac{}{10}$ $\frac{}{100}$

⑥ 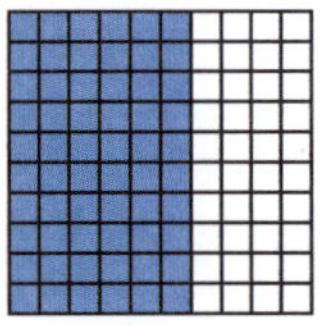 $\frac{}{10}$ $\frac{}{100}$

⑧ 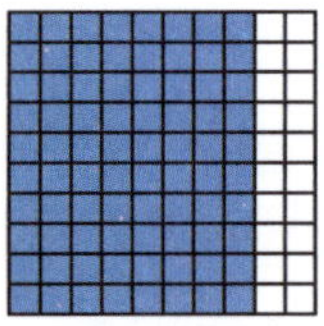 $\frac{}{10}$ $\frac{}{100}$

⑤ 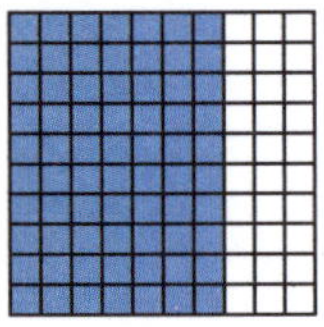 $\frac{}{10}$ $\frac{}{100}$

⑦ 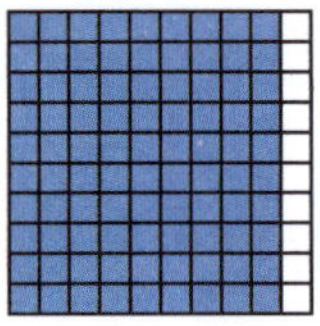 $\frac{}{10}$ $\frac{}{100}$

⑨ 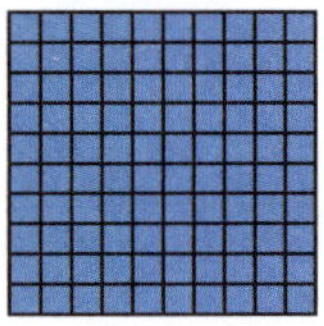 $\frac{}{10}$ $\frac{}{100}$

SCORE /9 0-4 5-7 8-9

Quick Check

Write these fractions in words as hundredths.

① $\frac{3}{10}$ thirty hundredths

② $\frac{5}{10}$ ______ hundredths

③ $\frac{7}{10}$ ______ hundredths

④ $\frac{1}{10}$ ______ hundredths

⑤ $\frac{11}{10}$ ______ hundredths

Write these hundredths in words as tenths.

⑥ $\frac{50}{100}$ ______ tenths

⑦ $\frac{80}{100}$ ______ tenths

⑧ $\frac{10}{100}$ ______ tenths

⑨ $\frac{100}{100}$ ______ tenths

⑩ $\frac{120}{100}$ ______ tenths

SCORE /10 0-4 5-8 9-10

TARGETING MATHS HOMEWORK: YEAR 4 © PASCAL PRESS ISBN: 9781925726572

Number & Algebra

AC9M4A02

The Area Model for Multiplication

To multiply 36 by 4, we can split 36 into 30 + 6 and then multiply.
We use a diagram called the **area model**.

×	30 +	6	
4	120	24	= 144

Complete the area models for these multiplications.

① 26 × 6

×	20 +	6	
6			= ______

② 27 × 3

×			
			= ______

③ 38 × 4

×			
			= ______

④ 72 × 3

×			
			= ______

⑤ 124 × 4

×	100 +	___ +	___	
				= ______

⑥ 253 × 6

×				
				= ______

SCORE /6 0-2 3-4 5-6

Statistics & Probability

AC9M4ST02

Take Care

You need to take care when you read or make a graph.
This is a graph that Kai made after a survey of the vegetables people preferred.

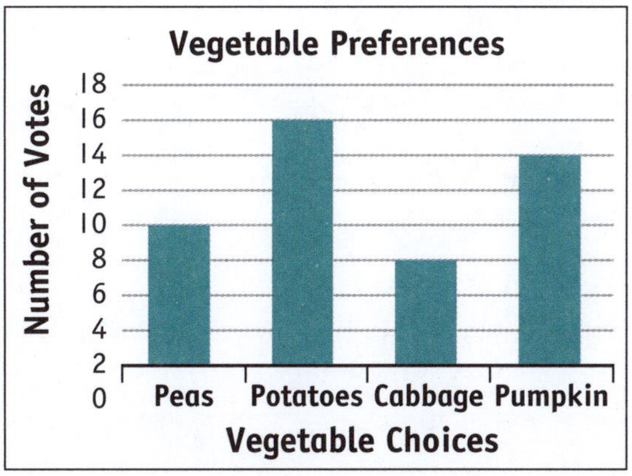

① How many people preferred each vegetable?

peas ___, potatoes ___, cabbage ___, pumpkin ___

② How many people did Kai survey? _____

The scale is incorrect as only 5 people preferred peas and 8 people preferred potatoes.

③ Redraw the graph with the correct scale.

④ How many people preferred each vegetable?

peas ___, potatoes ___, cabbage ___, pumpkin ___

⑤ How many people did Kai actually survey? _____

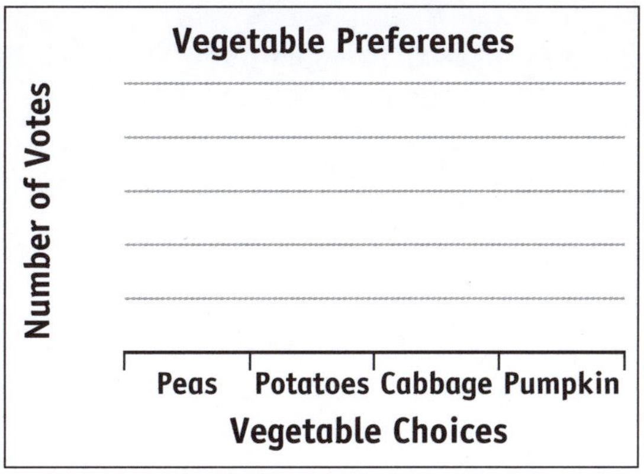

SCORE /5 0-2 3-4 5

Reading Temperature

Hot or Cold?

We use the Celsius scale to tell us how hot or cold something is.

In the Celsius scale, the temperature of water when it comes to the boil is 100 degrees, which is written as 100 °C. The temperature of water just as it freezes is 0 degrees Celsius, written as 0 °C.

Write the temperature shown on each thermometer.

① ______ ② ______ ③ ______ ④ ______ ⑤ ______

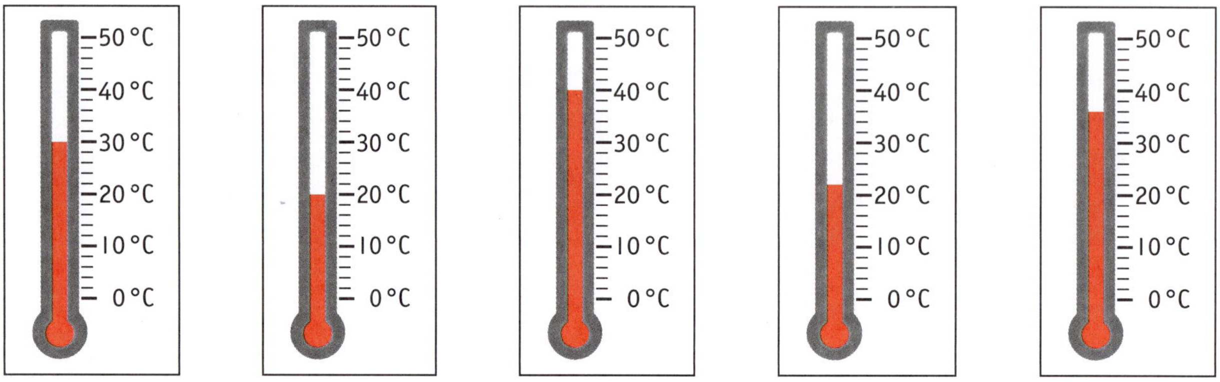

Use the numbers 1 to 5 to answer these questions.

⑥ Which thermometer shows the hottest temperature? ___

⑦ Which thermometer shows the coldest temperature? ___

SCORE /7 0-3 4-5 6-7

NAPLAN*-style Questions

① What decimal fraction does this diagram show?

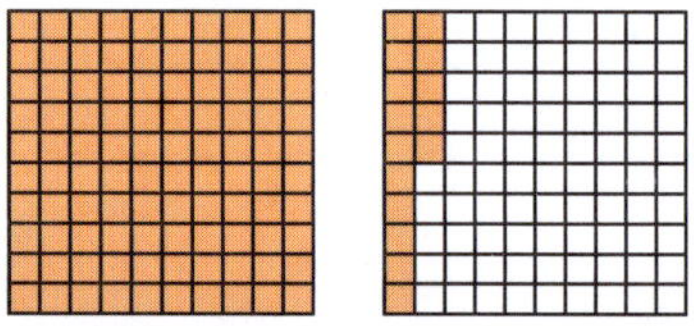

② Look at the graph.
What is the difference in height between bean 1 and bean 2?

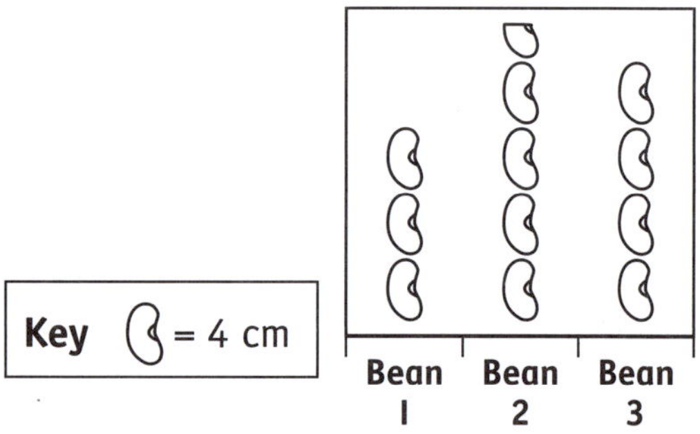

◯ 1·5 cm ◯ 3 cm ◯ 4·5 cm ◯ 6 cm

③ What is the temperature difference between the shade and the sunshine?

____ °C

In the shade

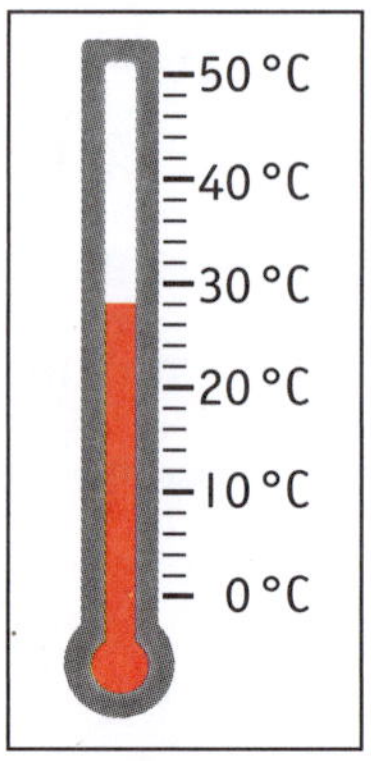

In the sunshine

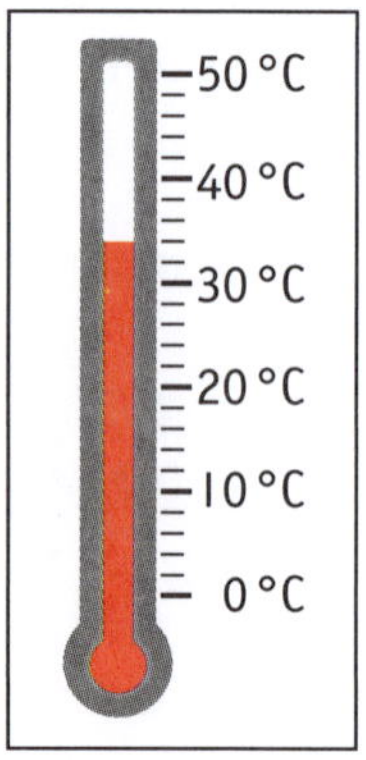

SCORE /3 0-1 2 3

TARGETING MATHS HOMEWORK: YEAR 4 © PASCAL PRESS ISBN: 9781925726572

Problem of the Week

AC9M4A02

Invent a Rule

"For this problem, I want you to invent a rule for always making the largest and smallest products possible with three digits that are made into a 1-digit number and a 2-digit number."

Mrs Green went on to tell her class that they should start with the digits 2, 3 and 4 and write all the possible 2-digit by 1-digit multiplications.

Then use the area model to work them out.

"For a start, you could use the area model as follows," Mrs Green said. She drew this on the board to get her class going:

× 20 + 3

4 [ ] [ ] = ______

Join Mrs Green's class. Use these area model diagrams to find the largest and smallest possible product with the digits 2, 3 and 4.

× [ ] [ ] = ______   × [ ] [ ] = ______   × [ ] [ ] = ______

× [ ] [ ] = ______   × [ ] [ ] = ______   × [ ] [ ] = ______

1. What is the largest product that you found? ________
2. What is the smallest product that you found? ________
3. Why are those the largest and smallest products?

____

____

Test the rule that you have invented by finding the largest and smallest products that can be made using the digits 3, 4 and 5.

× [ ] [ ] = ______   × [ ] [ ] = ______   × [ ] [ ] = ______

× [ ] [ ] = ______   × [ ] [ ] = ______   × [ ] [ ] = ______

If your theory worked again, test it one more time using three digits of your choice.

4. The digits chosen were __, __ and __.
5. The rule indicated that ____ × __ would make the largest product and that ____ × __ would make the smallest product.
6. Was your theory correct? _______

Decimal Notation

This diagram shows the decimal number 1·34.

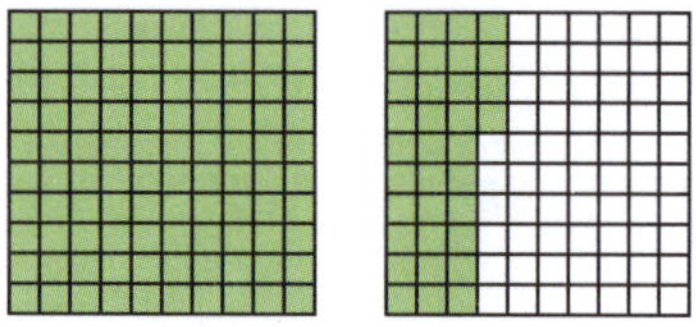

1·34 or $1\frac{34}{100}$

This diagram shows the decimal number 1·04.

1·04 or $1\frac{4}{100}$

Label these diagrams.

1. 

   __·____ or 1 ——

2. 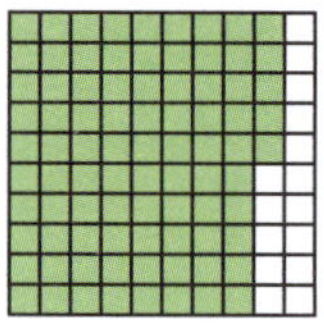

   __·____ or ——

3. 

   __·____ or __ ——

4. 

   __·____ or 1 ——

5. 

   __·____ or 1 ——

6. 

   __·____ or __ ——

SCORE /6 0-2 3-4 5-6

Quick Check

Write these decimals as fractions.

1. 1·3 $1\frac{3}{10}$ or $\frac{13}{\quad}$
2. 1·35 —— or ——
3. 1·03 —— or ——
4. 0·76 —— or ——
5. 2·38 —— or ——

Write these fractions as decimals.

6. $\frac{7}{10}$ __·____
7. $\frac{15}{10}$ __·____
8. $\frac{7}{100}$ __·____
9. $\frac{27}{10}$ __·____
10. $\frac{16}{100}$ __·____

SCORE /10 0-4 5-8 9-10

TERM 2

TARGETING MATHS HOMEWORK: YEAR 4 © PASCAL PRESS ISBN: 9781925726572

AC9M4N06

Number Splitting for Division

Dividing by a single digit can be made easier if you use number splitting.
If you want to divide 42 by 3 and you know that 3 × 10 = 30, then you can use that fact for the division.

For 42 ÷ 3, think 30 ÷ 3 = 10.

42 ÷ 3
30 + 12
÷ 3
10 + 4 = 14

For 56 ÷ 4, think 40 ÷ 4 = 10.

56 ÷ 4
40 + 16
÷ 4
10 + 4 = 14

Use number splitting for these divisions.

1. 51 ÷ 3 (Think 30 ÷ 3 = 10)
   30 + 21
   ÷ 3
   ___ + ___ = ___

2. 64 ÷ 4 (Think 40 ÷ 4 = 10)
   ___ + ___
   ÷ 4
   ___ + ___ = ___

3. 72 ÷ 6 (Think 60 ÷ 6 = 10)
   ___ + ___
   ÷ 6
   ___ + ___ = ___

4. 84 ÷ 7 (Think 70 ÷ 7 = 10)
   ___ + ___
   ÷ 7
   ___ + ___ = ___

5. 96 ÷ 8 (Think 80 ÷ 8 = 10)
   ___ + ___
   ÷ 8
   ___ + ___ = ___

6. 108 ÷ 9 (Think 90 ÷ 9 = 10)
   ___ + ___
   ÷ 9
   ___ + ___ = ___

SCORE /6 0-2 3-4 5-6

Statistics & Probability

AC9M4P01

Just by Chance

Jake shuffled his pack of cards and then took out one card. It was a spade. He put the card back in the pack, shuffled again and took out one card. It was a club. He did this three more times and every time the card he took out was black (spade or club).

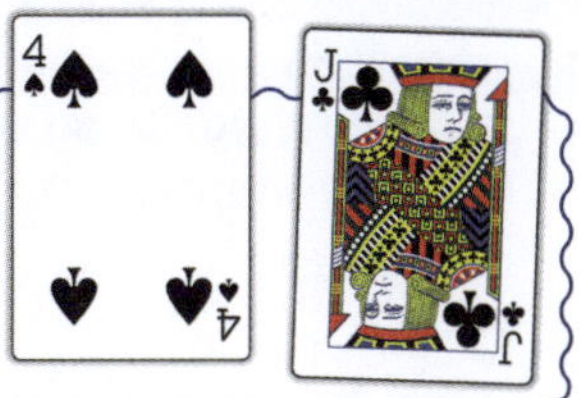

1. If Jake takes out one more card, what is the chance that it will be red (heart or diamond)?
   ◯ Impossible ◯ Unlikely ◯ Likely ◯ Very likely ◯ Certain

2. Explain your reasoning. ____________________

SCORE /2 0 1 2

AC9M4M01

Decimal Measurement

These lists show the measures we use for mass, capacity and length, and their abbreviations.

1000 grams = 1 kilogram	1500 g = 1·5 kg
1000 millilitres = 1 litre	500 mL = 0·5 L
100 centimetres = 1 metre	125 cm = 1·25 m
1000 metres = 1 kilometre	1250 m = 1·25 km

Complete these decimal measurement conversions.

1. 1350 g = ________
2. 2220 mL = ________
3. 1200 m = ________
4. 130 cm = ________
5. 350 cm = ________
6. 1·65 kg = ________
7. 3·4 m = ________
8. 1·7 L = ________
9. 1·06 kg = ________
10. 3·4 L = ________

11. 2650 g = ________
12. 763 mL = ________
13. 1200 m = ________
14. 2140 cm = ________
15. 980 cm = ________
16. 3·42 kg = ________
17. 1·04 m = ________
18. 52·6 L = ________
19. 0·04 kg = ________
20. 21·7 L = ________

SCORE /20 (0-9) (10-18) (19-20)

NAPLAN*-style Questions

1. What is the decimal number marked on this line?

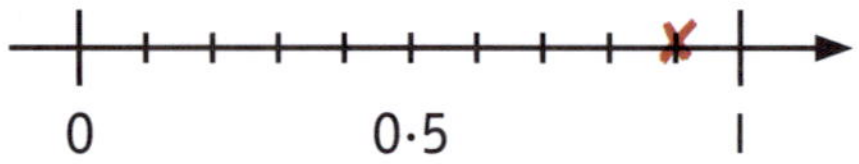

2. Claire wants 18 paper plates for a party. Plates are sold in packets of 6. How many packets of plates does Claire need to buy?

   ◯ 18 ◯ 6 ◯ 3 ◯ 24

3. Jake poured half a litre of juice into a container. Which of the options shows the amount he poured?

   ◯ 50 mL ◯ 500 g ◯ 500 mL ◯ 1000 mL

SCORE /3 (0-1) (2) (3)

TARGETING MATHS HOMEWORK: YEAR 4 © PASCAL PRESS ISBN: 9781925726572

Problem of the Week

Making a Flowchart

Kai is playing a guessing game with Jo. He has placed these cards on the table and chosen one without telling Jo.

Jo can ask questions to find out which card Kai chose.

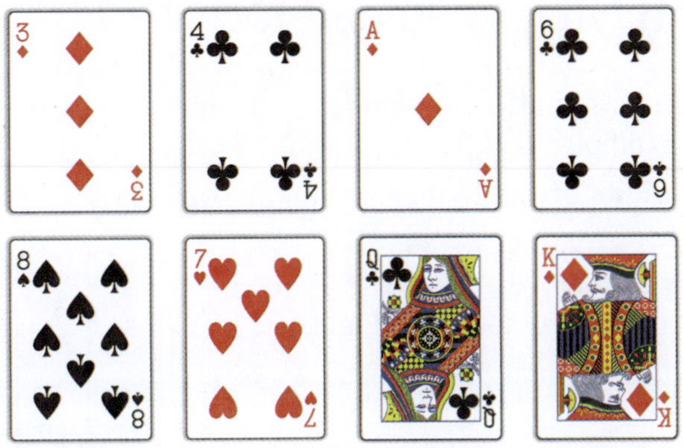

Follow the questions and flowchart to find out which card Kai chose.

These are Jo's questions:

1. Is it a black card?

2. Is it a heart?

3. Is it a picture card?

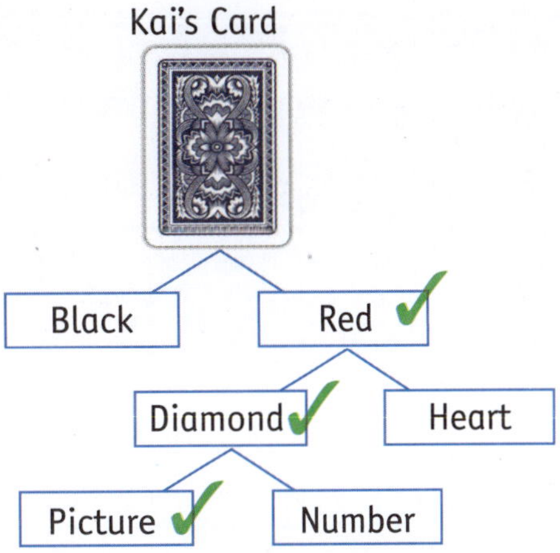

These are Kai's answers:

1. No

2. No

3. Yes

(1) Which card did Kai choose? ______________________

Try playing this game with a partner.

Record the questions and answers in the table and mark the progress on the flowchart.

Questions	Answers

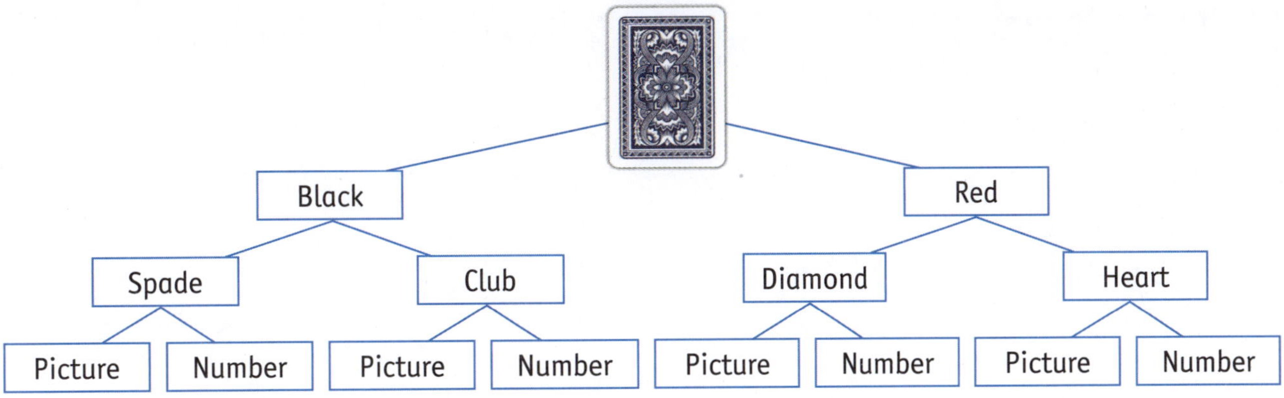

TARGETING MATHS HOMEWORK: YEAR 4 © PASCAL PRESS ISBN: 9781925726572

AC9M4N05

Multiplying by 10 and 100

When you multiply a number by 10 it becomes 10 times larger.
For example: 1 × 10 = 10 and 0·1 × 10 = 1.

The digits of the number being multiplied by 10 change position on a **place value** chart.

In this example, we see that 10 × 23·5 moves the digits 2, 3 and 5 one place to the left.

But the decimal point does not move – it separates the units from the tenths in both place value charts.

	Hundreds	Tens	Units	Tenths	Hundredths		Hundreds	Tens	Units	Tenths	Hundredths
10 ×		2	3 •	5		becomes	2	3	5 •		

TERM 2

Show these on the place value chart using the **move one place left rule** for multiplying by 10.

		H	T	U	T	H		H	T	U	T	H
① Show 1·1 × 10	10 ×			1 •	1		becomes			•		
② Show 10 × 0·01	10 ×			•						•		
③ Show 3·46 × 10	10 ×			•						•		
④ Show 82·3 × 10	10 ×			•						•		

When you multiply by 100, every digit moves two places to the left.
Complete these questions using the **move two places left rule** for multiplying by 100.

		H	T	U	T	H		H	T	U	T	H
⑤ Show 100 × 1	100 ×			1 •			becomes			•		
⑥ Show 100 × 0·11	100 ×			•						•		
⑦ Show 100 × 1·34	100 ×			•						•		
⑧ Show 100 × 7·06	100 ×			•						•		

SCORE /8 0-3 4-6 7-8

Quick Check

Multiply these numbers by 10.

① 0·8 becomes ______.

② 0·75 becomes ______.

③ 1·3 becomes ______.

④ 1·03 becomes ______.

⑤ 2·57 becomes ______.

Multiply these numbers by 100.

⑥ 0·8 becomes ______.

⑦ 0·75 becomes ______.

⑧ 1·3 becomes ______.

⑨ 1·03 becomes ______.

⑩ 2·57 becomes ______.

SCORE /10 0-4 5-8 9-10

TARGETING MATHS HOMEWORK: YEAR 4 © PASCAL PRESS ISBN: 9781925726572

AC9M4N08

Rounding Money Amounts

When we go shopping, we can round money amounts to the nearest 10c to make a good estimation of the amount that we have spent.

As with rounding numbers, amounts with 4c or less round down to the dollar below, and amounts of 5c or more round up to the next dollar.

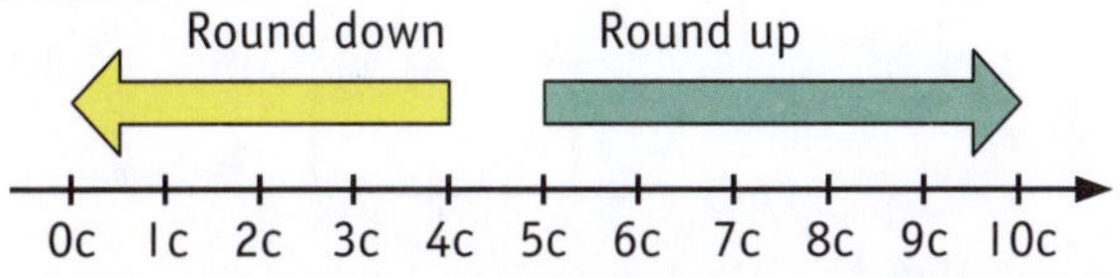

Complete the rounded prices and then add.

①

	Item price	Rounded price
	$1·62	$ 1·60
	$2·35	$ 2·40
	$3·99	$ ______
	$4·15	$ ______
Total	$ ______	$ ______

②

	Item price	Rounded price
	$1·75	$ ______
	$2·95	$ ______
	$1·15	$ ______
	$2·03	$ ______
Total	$ ______	$ ______

③

	Item price	Rounded price
	$3·55	$ ______
	$4·64	$ ______
	$2·23	$ ______
	$5·12	$ ______
Total	$ ______	$ ______

④

	Item price	Rounded price
	$3·45	$ ______
	$2·19	$ ______
	$3·94	$ ______
	$1·22	$ ______
Total	$ ______	$ ______

SCORE /4 0-1 2-3 4

Statistics & Probability

AC9M4ST02

A Sales Comparison

The Craft Publishing Company compared their 2022 sales with their 2023 sales in this chart. **Note:** Sales figures were rounded to the nearest 5000 books.

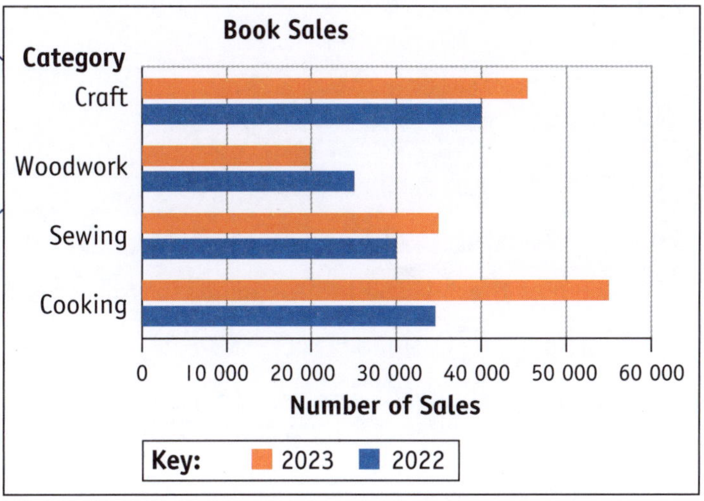

① Which category sold the most in 2022?

____

② Which category sold the most in 2023?

____

③ Which category grew the most?

____

④ By how much did Woodwork sales decrease between 2022 and 2023? ______________

SCORE /4 0-1 2-3 4

TARGETING MATHS HOMEWORK: YEAR 4 © PASCAL PRESS ISBN: 9781925726572

AC9M4SP01

Making Characters with Shapes

Zoe and her class have been given a brief to make characters for the new cartoon movie *Shapeland*. The only shapes they can use are:

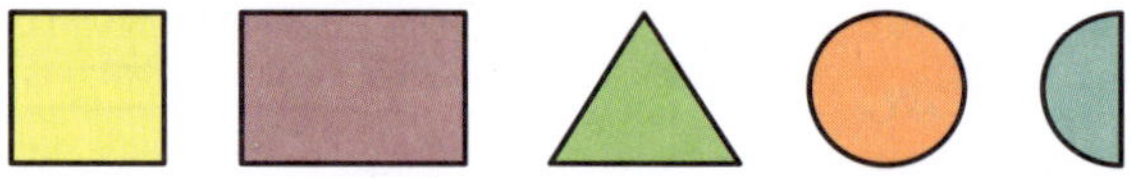

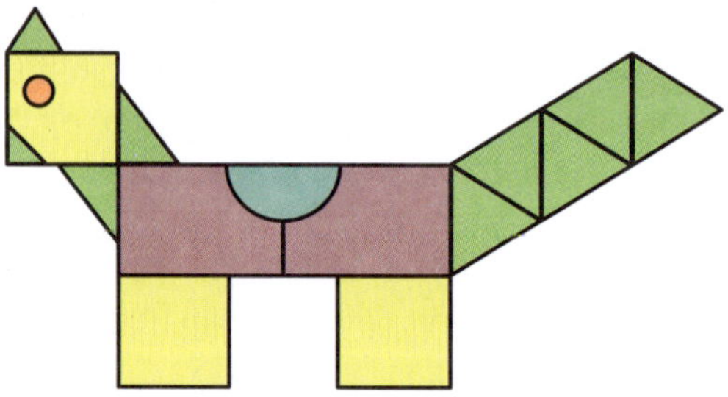

9 triangles
1 circle
1 semicircle
2 rectangles
3 squares

Become a designer for the new movie Shapeland and design and label two more characters.

SCORE /2 (0) (1) (2)

NAPLAN*-style Questions

1. Which decimal number means $\frac{175}{100}$?

   ◯ 175 ◯ 17·5 ◯ 1·75 ◯ 0·175

2. Which of these is the best estimate for 6 pencils at $1·54 each?

   ◯ $6 ◯ $9·30 ◯ $9·27 ◯ $9

3. How many triangles were used to make this shape?

   ▭

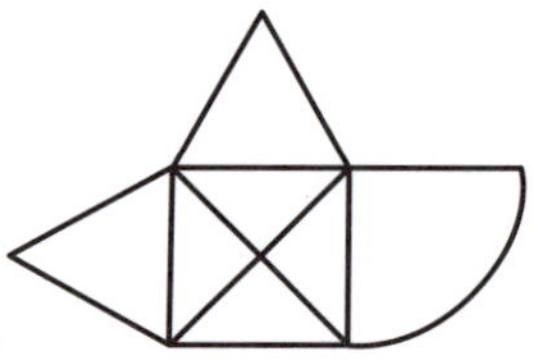

SCORE /3 (0-1) (2) (3)

TARGETING MATHS HOMEWORK: YEAR 4 © PASCAL PRESS ISBN: 9781925726572

Screen Time

The Servin family decided to collect data about their screen time between 4:00 pm, when their daughters arrive back from school, and 7:00 pm when the family have their evening meal.

They showed their data in two ways: on a time chart (Graph 1) and a line graph (Graph 2).

Graph 1

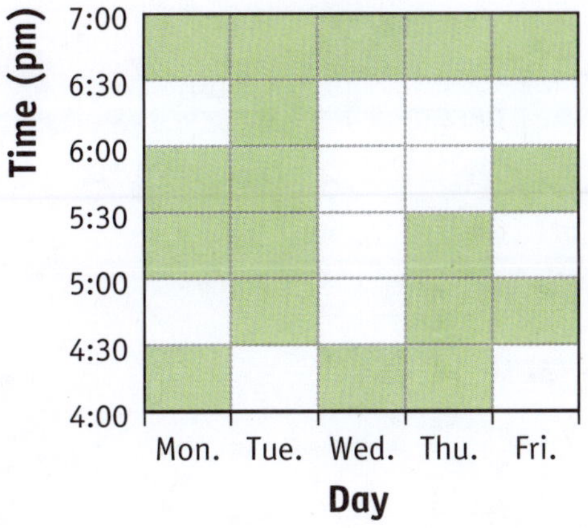

Graph 2

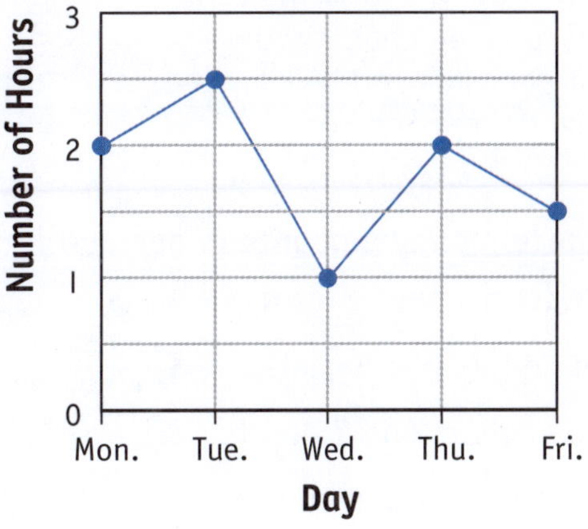

Use the graphs to answer these questions.
Say whether you found Graph 1 or Graph 2 easier to use to answer that question.

1. During which half hour do the Servins have the most screen time?

   Between ___ and ___.

   I used Graph __ for this question because ____.

2. On which day of the week do the Servins have $1\frac{1}{2}$ hours of screen time during the week?

   On _______________.

   I used Graph __ for this question because ____.

3. On which day of the week do the Servins have least screen time?

   On _______________.

   I used Graph __ for this question because ____.

4. On which day of the week do the Servins have most screen time?

   On _______________.

   I used Graph __ for this question because ____.

Collect data on screen time in your house and show your data on these two graphs.

Graph 1

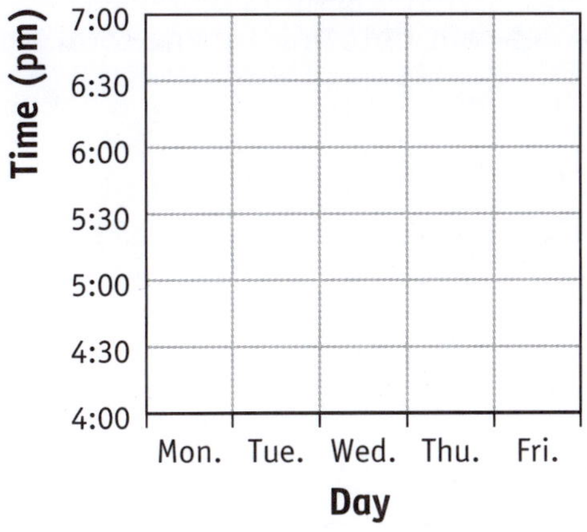

Graph 2

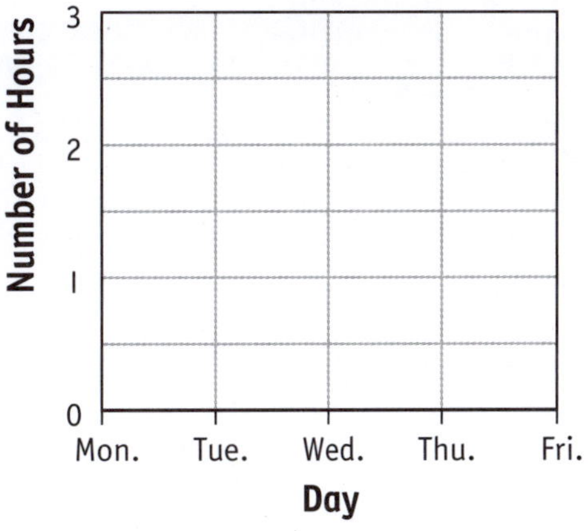

Decimal Notation and Money

Money is written using decimal notation.

One dollar and fifteen cents is written using decimals as $1·15. The decimal point separates the whole dollars from the cents. The place value chart that we use for decimals can be changed to show dollars and cents.

Hundreds	Tens	Units	Tenths	Hundredths
		1 •	1	5

becomes

Dollars			Cents	
H	T	U	T	H
		1 •	1	5

Write these money amounts in decimal form.

1. Two dollars and 35 cents $ __·_____
2. Five dollars and 5 cents $ __·_____
3. Three dollars and 75 cents $ __·_____
4. Twelve dollars and 10 cents $ _____·_____
5. 153 cents $ __·_____
6. 205 cents $ __·_____
7. 1400 cents $ _____·_____
8. 2345 cents $ _____·_____

Adding Cents

When we add amounts given in cents, we usually change the total to a **dollar-and-cent** amount.

If we bought an item for 50 cents and another for 60 cents, the total amount would be 50c + 60c = 110c. We change this to decimal form, $1·10.

Note that when writing money amounts, we keep the zero in the cents part because $1·1 might be read as 'one dollar and one cent' when the actual value is 'one dollar and ten cents'.

Add these amounts and show the total in dollar-and-cent notation.

9. 75c + 30c = $__·_____
10. 80c + 90c = $__·_____
11. 85c + 30c = $__·_____
12. $1·95 + 25c = $__·_____
13. $3·75 + $1·35 = $__·_____
14. $7·35 + $2·25 = $__·_____
15. $6·45 + $4·65 = $_____·_____
16. $12·43 + $9·57 = $_____·_____

SCORE /16 0-7 8-14 15-16

Quick Check

Write these amounts in dollar-and-cent notation.

1. 155 cents = $__·_____
2. 350 cents = $__·_____
3. 305 cents = $__·_____
4. 125 cents = $__·_____
5. 245 cents = $__·_____

Add these money amounts and use dollar-and-cent notation for the total.

6. 75c + 28c = $__·_____
7. 36c + 74c = $__·_____
8. 125c + 85c = $__·_____
9. 220c + 80c = $__·_____
10. $1·50 + 55c = $__·_____

SCORE /10 0-4 5-8 9-10

TARGETING MATHS HOMEWORK: YEAR 4 © PASCAL PRESS ISBN: 9781925726572

Number & Algebra

UNIT 13

AC9M4N06, AC9M4A02

Multiplication and Division Fact Families

Multiplication facts can be used to help with division because they belong to the same family. When you know that 4 × 6 = 24, it helps with 24 ÷ 4 = 6 and 24 ÷ 6 = 4.

Write the related division facts for these multiplications.

1. Use 7 × 5 = ☐ for 35 ÷ 7 = ☐ and 35 ÷ 5 = ☐
2. Use 8 × 4 = ☐ for ☐ ÷ 8 = ☐ and ☐ ÷ 4 = ☐
3. Use 6 × 9 = ☐ for ☐ ÷ ☐ = ☐ and ☐ ÷ ☐ = ☐
4. Use 9 × 7 = ☐ for ☐ ÷ ☐ = ☐ and ☐ ÷ ☐ = ☐
5. Use 4 × 9 = ☐ for ☐ ÷ ☐ = ☐ and ☐ ÷ ☐ = ☐
6. Use 7 × 6 = ☐ for ☐ ÷ ☐ = ☐ and ☐ ÷ ☐ = ☐
7. Use 8 × 6 = ☐ for ☐ ÷ ☐ = ☐ and ☐ ÷ ☐ = ☐
8. Use 7 × 3 = ☐ for ☐ ÷ ☐ = ☐ and ☐ ÷ ☐ = ☐
9. Use 3 × 12 = ☐ for ☐ ÷ ☐ = ☐ and ☐ ÷ ☐ = ☐
10. Use 8 × 9 = ☐ for ☐ ÷ ☐ = ☐ and ☐ ÷ ☐ = ☐

SCORE /10 0-4 5-8 9-10

TERM 2

Statistics & Probability

AC9M4P01

How Well Do You Know Your 9 Times Table?

You are going to say your 9 times table and score 1 point for every 9 times fact that you get right. But before you start, rate your chance of getting each of these scores using the chance words **impossible**, **unlikely**, **likely**, **very likely** and **certain**.

Once you have made your predictions, say the 9 times table to a partner. Your partner records what you say so that you can check how many you get right.

Score out of 10	Prediction	How close was your prediction?
Up to 3		
4 or 5		
6, 7 or 8		
9 or 10		

- Now practise the facts that you didn't know.
- If you already know your 9 times facts, try the experiment with the 13 times table.
- Make new predictions and test them.

Score out of 10	Prediction	How close was your prediction?
Up to 3		
4 or 5		
6, 7 or 8		
9 or 10		

Decimals for Measuring

Measurements of length are written using decimal notation.

One metre and fifteen centimetres is written using decimals as 1·15 m.

The decimal point separates the whole metres from the centimetres.

The place value chart that we use for decimals can be changed to show metres and centimetres.

Hundreds	Tens	Units	Tenths	Hundredths
		1 •	1	5

becomes

Metres			Centimetres	
H	T	U	T	H
		1 •	1	5

Write these measurements in decimal form.

1. 135 cm = ____ · ______ m
2. 105 cm = ____ · ______ m
3. 330 cm = ____ · ______ m
4. 250 cm = ____ · ______ m
5. 4700 cm = ____ · ______ m
6. 12 cm = ____ · ______ m
7. Two metres and thirty-five centimetres = ____ · ______ m
8. One metre and sixty-four centimetres = ____ · ______ m
9. Four metres and five centimetres = ____ · ______ m
10. Seven metres and sixty centimetres = ____ · ______ m
11. Twelve metres and forty centimetres = ____ · ______ m
12. Eighty-two centimetres = ____ · ______ m

SCORE /12 0-5 6-10 11-12

NAPLAN*-style Questions

1. How many packets of 6 sausage rolls can you make by cooking 2 trays of 24 sausage rolls?

   ◯ 24 ◯ 6 ◯ 12 ◯ 8

2. Jake bought 3 pencils that cost $1·45 each.
   How much money did Jake spend?

   $ ________

3. What is the chance of pulling a red counter out of the bag without looking?

   ◯ Impossible ◯ Unlikely ◯ Likely
   ◯ Very likely ◯ Certain

SCORE /3 0-1 2 3

TARGETING MATHS HOMEWORK: YEAR 4 © PASCAL PRESS ISBN: 9781925726572

Problem of the Week

AC9M4N06, AC9M4A02

A Minute Game

We really enjoy playing the Minute Game.

The aim of this game is to make a number that is as close as possible to the **target number**. These are the rules that we made up.

- Roll one die: 
- Then roll another die: 
- This means that the **target number** is **14**.
- Roll three more dice:  4, 3 and 6 are the digits that we have to work with.
- We set a timer to 1 minute.
- In that time, we have to make a total that is as close to the target as possible, using any operations or combinations of the digits.
- At the end of one minute, the player closest to the target number wins 1 point. If there is a draw, each player wins half a point.

Here is the result of our first game.

I thought that $6 + 4 + 3 = 13$ would win.

My combination was $6 \times 3 - 4 = 14$ and that did win!

I thought that $(6 - 3) \times 4 = 12$ would win.

Try these examples. Remember, you may not be able to make the target exactly. Set your timer and then go!

1. **Target:** 23 **Dice:** 

Working space	Closest:

2. **Target:** 34 **Dice:** 

Working space	Closest:

3. Try a game of your own making.

Target: ___ **Dice:**

Working space	Closest:

AC9M4N02, AC9M4N09

Odd and Even Numbers

I entered this key sequence into my calculator:

1 [+] 2 [=] [=] [=]

I thought I would see 2, 4, 6, 8 in the display, but all the totals were odd numbers.

I entered this key sequence into my calculator:

2 [+] 2 [=] [=] [=]

I saw 2, 4, 6, 8 in the display.
I had started with an even number and all my totals were even numbers.

"Is that always true?" they wondered.

Enter these key sequences into your calculator and complete the totals pattern.

Key sequence to enter	Odd–Even pattern in the totals
① 3 [+] 3 [=] [=] [=] [=] [=]	E, O, E, O, E, __, __
② 3 [+] 4 [=] [=] [=] [=] [=]	__, __, __, __, __, __, __
③ 2 [+] 3 [=] [=] [=] [=] [=]	__, __, __, __, __, __, __
④ 31 [+] 3 [=] [=] [=] [=] [=]	__, __, __, __, __, __, __
⑤ 25 [+] 3 [=] [=] [=] [=] [=]	__, __, __, __, __, __, __

What happens when you multiply? Will the same patterns emerge in the totals? Complete these sequences to find out.

⑥ 1 [×] 2 [=] [=] [=] [=] [=]	___, ___, ___, ___, ___, ___, ___
⑦ 2 [×] 3 [=] [=] [=] [=] [=]	___, ___, ___, ___, ___, ___, ___
⑧ 3 [×] 5 [=] [=] [=] [=] [=]	___, ___, ___, ___, ___, ___, ___
⑨ 6 [×] 4 [=] [=] [=] [=] [=]	___, ___, ___, ___, ___, ___, ___
⑩ 1 [×] 3 [=] [=] [=] [=] [=]	___, ___, ___, ___, ___, ___, ___

SCORE /10 0-4 5-8 9-10

Quick Check

Will the answer be odd or even?

① 2 + 4 + 3 ________

② odd + odd + even = ________

③ even + even + odd = ________

④ odd + odd + odd = ________

⑤ even + even + even = ________

⑥ 2 × 3 ________

⑦ 3 × 5 ________

⑧ odd × odd = ________

⑨ odd × even = ________

⑩ even × even = ________

SCORE /10 0-4 5-8 9-10

TARGETING MATHS HOMEWORK: YEAR 4 © PASCAL PRESS ISBN: 9781925726572

Equivalent Additions

We say that 2 + 4 is **equivalent** to 1 + 5 because both additions have the same total:
2 + 4 = 6 and **1 + 5 = 6.**
You can think of this as 2 + 4 balances 1 + 5.

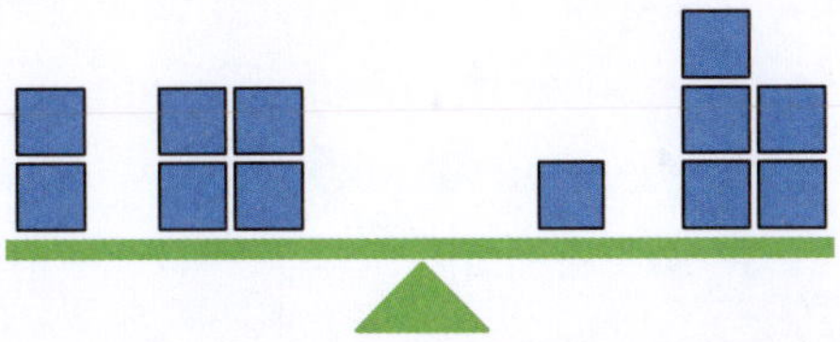

Complete these equivalent additions.

1. 7 + 7 = 9 + 5
2. 8 + 9 = 6 + ___
3. 27 + 6 = 17 + ___
4. 33 + 3 = 30 + ___

When you see the patterns in these equivalent additions, you can work smart not hard:

7 + 7 = 9 + 5
(7 + 2) + (7 − 2) = 9 + 5

8 + 9 = 6 + 11
(8 − 2) + (9 + 2) = 6 + 11

See if you can work smart, not hard, to make these additions equivalent.

5. 18 + 6 = 20 + ___
6. 27 + 5 = 17 + ___
7. 29 + 9 = 30 + ___
8. 36 + 12 = 26 + ___
9. 75 + 15 = 80 + ___
10. 19 + 13 = 29 + ___

SCORE /10 0-4 5-8 9-10

Statistics & Probability

AC9M4P01

What Chance of Going to the Movies?

1. "I am sure that I will be going to the movies tomorrow."
How would you rate Claire's chance of going to the movies tomorrow?
◯ Impossible ◯ Unlikely ◯ Likely ◯ Very likely ◯ Certain

2. Claire's friend Chen rang to say that she would not be able to go with Claire tomorrow.
What do think Claire's chance of going to the movies tomorrow is now?
◯ Impossible ◯ Unlikely ◯ Likely ◯ Very likely ◯ Certain

3. The weather forecast now says that there will be severe storms tomorrow.
What do think Claire's chance of going to the movies tomorrow is now?
◯ Impossible ◯ Unlikely ◯ Likely ◯ Very likely ◯ Certain

4. Claire woke up with a temperature and a sore throat.
What do think Claire's chance of going to the movies tomorrow is now?
◯ Impossible ◯ Unlikely ◯ Likely ◯ Very likely ◯ Certain

5. Does the chance of something happening change over time? ______
Why do you think that is? ________________________________

SCORE /5 0-2 3-4 5

Measurement & Space

AC9M4SP02

Where is X?

X should mark the spot on the island where the treasure is buried, but where is X?

- X is not in the northern half of the map.
- X is not in an even-numbered row.
- X is not close to the sea.
- X is under a palm tree.

Mark the spot where the treasure is buried.

The Treasure Map

NAPLAN*-style Questions

1. Will the tenth number in this pattern sequence be odd or even?

   2, 5, 8, 11, ...

   ◯ Odd ◯ Even

2. Complete this equivalent addition.

   36 + ____ = 27 + 19

3. A coin was tossed eight times and the results were recorded. What is the chance that the next toss will be heads?

Heads	Tails
𝍸 ||	|

◯ Impossible ◯ Unlikely ◯ Likely ◯ Very likely ◯ Certain

SCORE /3 0-1 2 3

TARGETING MATHS HOMEWORK: YEAR 4 © PASCAL PRESS ISBN: 9781925726572

In Rows and Circles

We have been placing the numbers 2–10 in the small circles of this diagram.

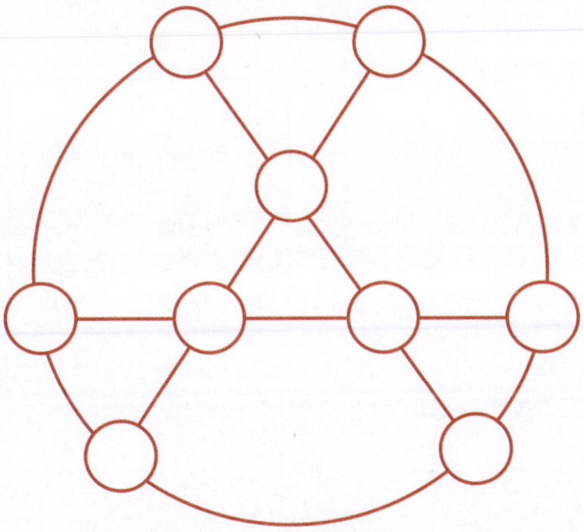

We made this number chart and ticked them off as we went to make sure that we used each number once and once only.

2	3	4	5	6	7	8	9	10

Kai and Jo now have some challenges for you.

Write the numbers 2–10 in the small circles of our diagram, above.

1. Find the total along each line.

   Line 1 Total = ____ Line 2 Total = ____ Line 3 Total = ____

2. Find the total around the outside circle.

   Circle Total = ____

3. Find a way of placing the numbers so that the totals along each line and around the circle are the same. Use these diagrams to help find your arrangements.

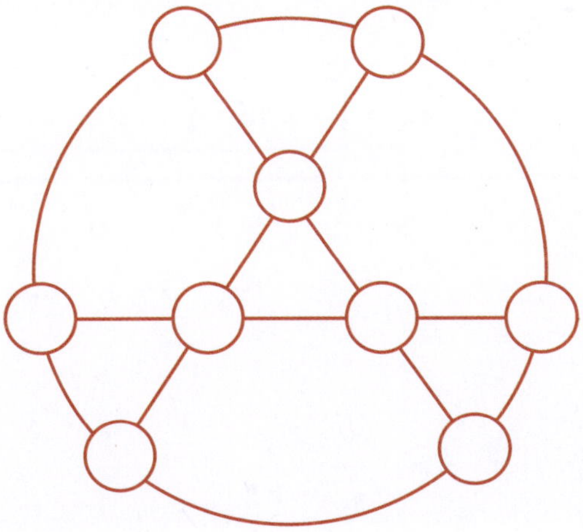

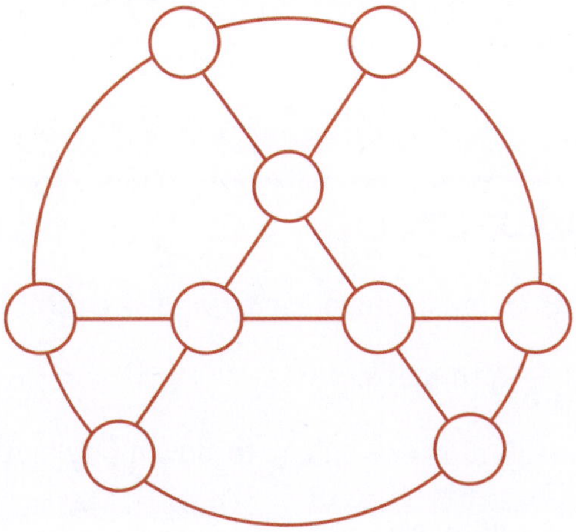

Hint: We found the last question hard until we worked out what that total should be. To do that, we added all the numbers 2–10.

Then, because each number will be used twice (on two lines or a line and a circle), we doubled that total.

Because there are three lines and one circle that all need the same total, we divided the doubled sum by 4 to find what the total along each line and around the circle should be.

TERM 2

AC9M4N03

Landmark Numbers

The numbers **25, 50, 75** and **100** are important **landmark numbers.**

2 × 50 = 100, 4 × 25 = 100, 3 × 25 = 75

The landmark numbers also appear in decimal fractions and in measurements.

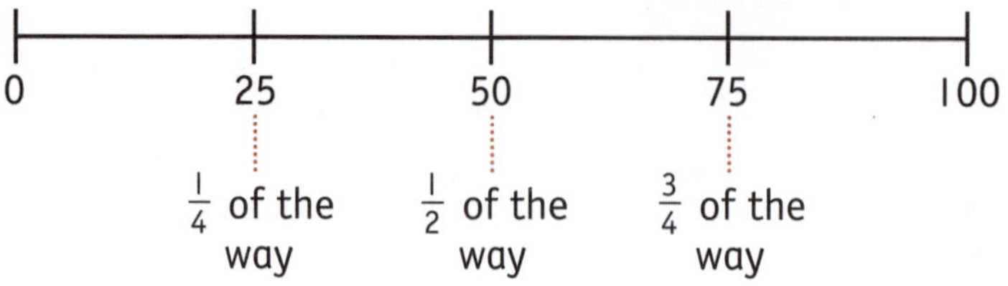

$\frac{1}{4} = \frac{25}{100} = 0{\cdot}25$

$\frac{1}{2} = \frac{50}{100} = 0{\cdot}5$

$\frac{3}{4} = \frac{75}{100} = 0{\cdot}75$

Complete these examples of landmark numbers.

1. $\frac{1}{2}$ of a metre is ______ cm or 0·______ m.
2. $\frac{1}{4}$ of a metre is ______ cm or 0·______ m.
3. $\frac{3}{4}$ of a metre is ______ cm or 0·______ m.
4. $1\frac{1}{2}$ metres is ______ cm or __·______ m.
5. $1\frac{1}{4}$ metres is ______ cm or __·______ m.
6. Two $\frac{1}{4}$ m lengths of elastic is ______ cm or __·______ m.
7. Three lengths of 0·5 m of elastic is ______ cm or __·______ m.
8. Three half metres of elastic is ______ cm or __·______ m.
9. One and three quarters of a metre is ______ cm or __·______ m.
10. Three lengths of 0·25 metres is ______ cm or __·______ m.

SCORE /10 0-4 5-8 9-10

Quick Check

1. 0·75 m is equivalent to what fraction of a metre?

   — of a metre

2. 1·25 m is equivalent to how many metres, as a fraction?

   — metres

3. How many centimetres is 0·5 m?

   _____ cm

4. What is $\frac{1}{4}$ m in decimal form?

   __ ·_____ metres.

5. 50 cm of elastic is the same as 0·__ m of elastic or — m of elastic.

SCORE /5 0-2 3-4 5

TARGETING MATHS HOMEWORK: YEAR 4 © PASCAL PRESS ISBN: 9781925726572

Number & Algebra

AC9M4N09, AC9M4A02

Multiplying by 9

There is an algorithm for **multiplying by 9** which means that you don't need to know all the × 9 facts.

For 2 × 9, you can think 2 × 10 − 2 = 18.

For 3 × 9, you can think 3 × 10 − 3 = 27.

For 9 × 9, you can think 9 × 10 − 9 = 81.

Use the × 9 strategy for these examples.

1. 6 × 9 = 6 × 10 − 6 = ___
2. 8 × 9 = ___ × ___ − ___ = ___
3. 13 × 9 = ___ × ___ − ___ = ___
4. 20 × 9 = ___ × ___ − ___ = ___
5. 16 × 9 = ___ × ___ − ___ = ___
6. 35 × 9 = ___ × ___ − ___ = ___
7. 19 × 9 = ___ × ___ − ___ = ___
8. 26 × 9 = ___ × ___ − ___ = ___
9. 23 × 9 = ___ × ___ − ___ = ___
10. 89 × 9 = ___ × ___ − ___ = ___

SCORE /10 0-4 5-8 9-10

Statistics & Probability

AC9M4ST03

Kitty's Growth Chart

Kitty has been sick, but she still needed to grow about 70 g in 7 days. We weighed Kitty at 8:00 am every day.

Here is her growth chart for the week.

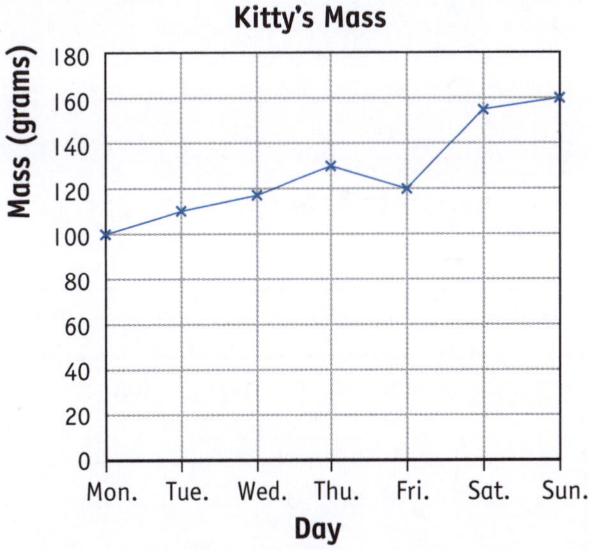

1. What did Kitty weigh on Monday morning?

   ________ g

2. On which day did Kitty's mass go down?

   ____

3. On which day did Kitty show the most gain?

   ____

4. On which day did Kitty show the least gain?

   ____

5. How many grams did Kitty gain over the week?

   ________ g

6. How far short of her growth target for the week was Kitty?

   ________ g

SCORE /6 0-2 3-4 5-6

AC9M4M02

Measuring Area

Area tells us how much space there is inside a shape.
It is usually measured in squares.

What is the area of each of these shapes?

① _____ squares

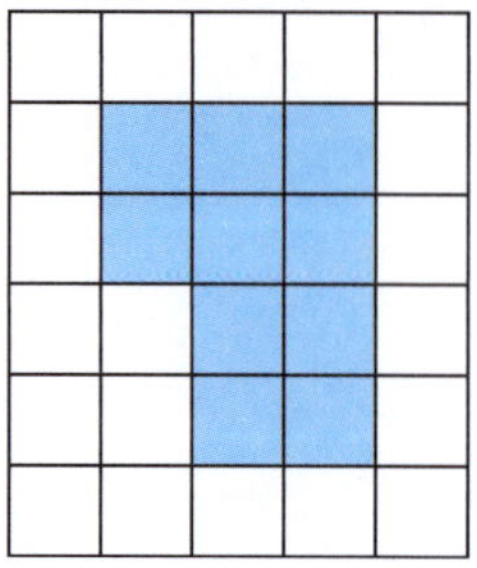

② _____ squares

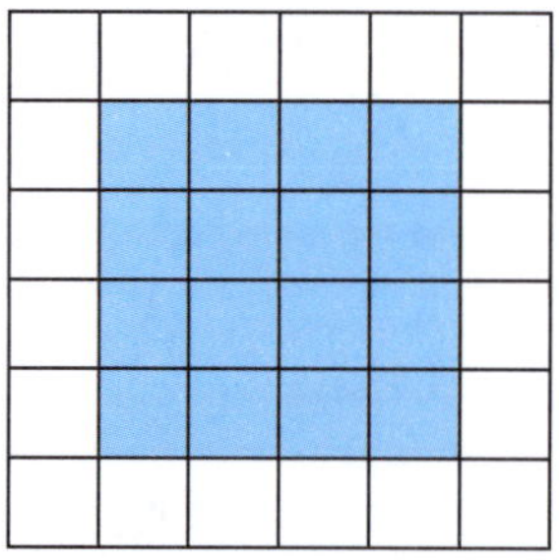

③ _____ squares

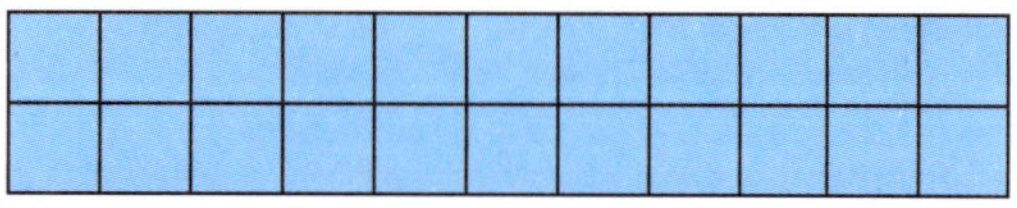

④ _____ squares

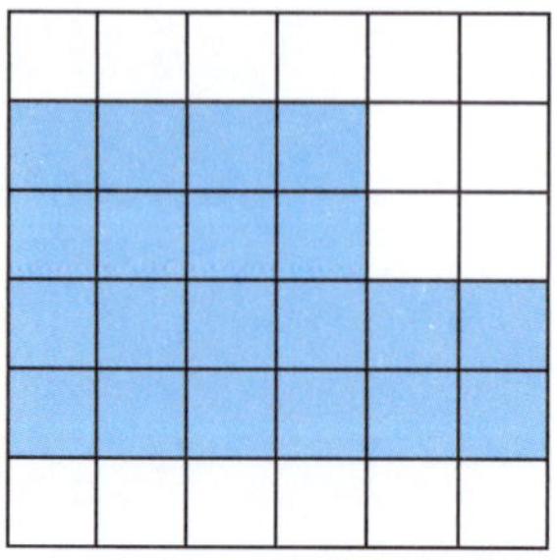

⑤ _____ squares

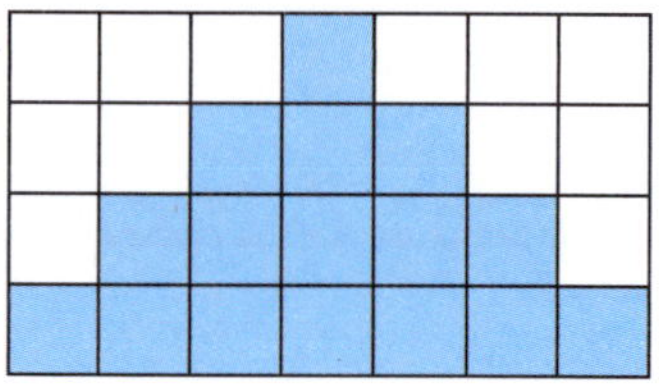

⑥ _____ squares

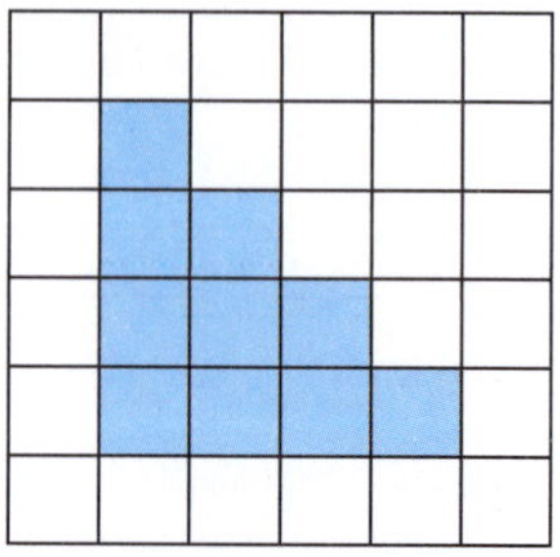

⑦ _____ squares

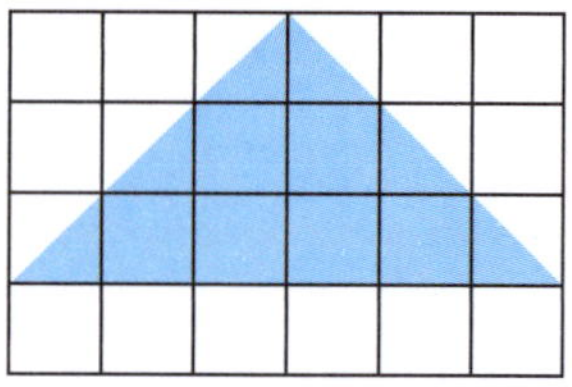

SCORE /7 0-3 4-5 6-7

NAPLAN*-style Questions

① 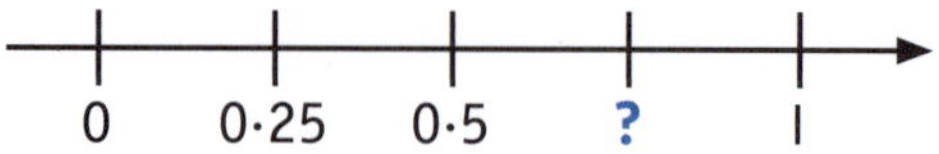

What decimal does the question mark represent? _______

② Which expression shows how to work out 12 × 9?

◯ 10 × 9 + 2 ◯ 12 × 10 − 9 ◯ 12 × 10 − 12 ◯ 12 × 9 − 2

③ What is the area of the green shape?

◯ 6 squares
◯ 8 squares
◯ 6·5 squares
◯ 7 squares

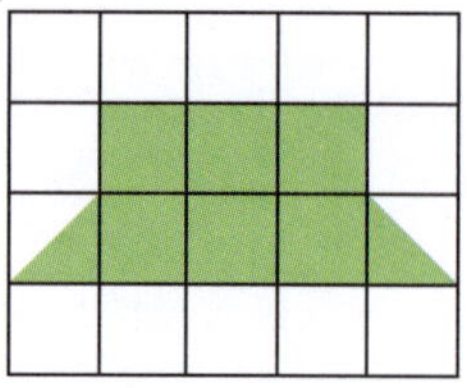

SCORE /3 0-1 2 3

TERM 2

TARGETING MATHS HOMEWORK: YEAR 4 © PASCAL PRESS ISBN: 9781925726572

The Farmer's Hurdles

A hurdle is a short piece of fencing that shepherds use to quickly make a pen for their sheep.

Hurdle

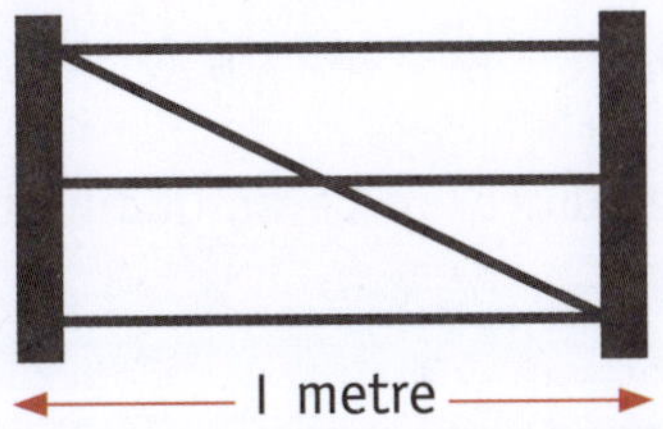

① How can you use 12 hurdles to enclose a sheep pen of exactly 6 square metres?

Use the grid at right to show your sheep pen.

② How many different-shaped pens (each enclosing 6 square metres) can you make with 12 hurdles?

Use the grid below to show your sheep pens.

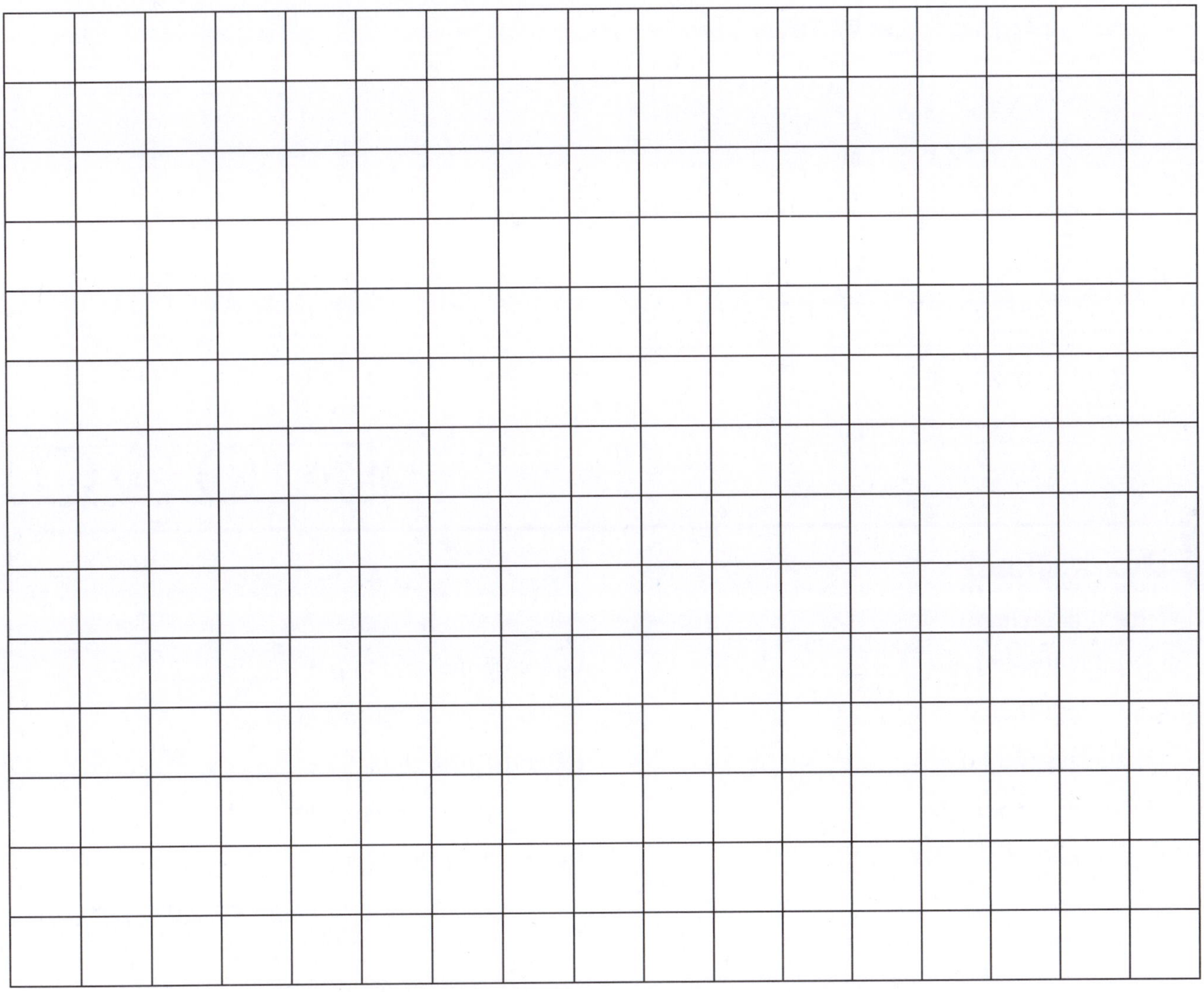

TERM 2

AC9M4N01

Counting in Decimals

For each of the sequences below, you will need to:

- decide whether the sequences are increasing or decreasing
- fill in the missing decimals
- set your calculator to be an adding or subtracting machine and use it to check your answers.

Here's an example:

0·9, 1·0, 1·1, __·__, __·__, 1·4, 1·5.

This is an increasing sequence and the missing decimals are 1·2 and 1·3.

To check, set your calculator to be an adding machine with this key sequence:

0·9 [+] 0·1 [=] [=] [=] [=]

To check that you have the right sequence, keep pressing the = button and the calculator adds 0·1 to the display every time.

Complete these counting in decimals sequences and check with your calculator.

	Decimal Sequence	Calculator Key Sequence
1	3·3, 3·1, __·__, __·__, 2·5, 2·3 Decreasing by ___	3·3 [−] 0·2 [=] [=] [=]
2	2·8, 3·1, 3·4, __·__, __·__, 4·3 ________ by ___	____________
3	4·2, 3·7, 3·2, __·__, __·__, 1·7 ________ by ___	____________
4	5, 5·25, 5·5, __·__, __·__, __·__ ________ by ___	____________
5	6·1, 7·2, 8·3, __·__, __·__, __·__ ________ by ___	____________

SCORE /5 (0-2) (3-4) (5)

Quick Check

Write the decimal.

1. 0·2 more than 5·7 __·__
2. 0·3 less than 3·2 __·__
3. 0·5 less than 0·6 __·__
4. 0·5 more than 2·2 __·__
5. 1·2 more than 3·03 __·__
6. 0·05 more than 0·8 __·__
7. 0·15 more thatn 1·6 __·__
8. 0·01 less than 1·23 __·__
9. 0·01 more than 0·99 __·__
10. 0·1 less than 0·95 __·__

SCORE /10 (0-4) (5-8) (9-10)

TARGETING MATHS HOMEWORK: YEAR 4 © PASCAL PRESS ISBN: 9781925726572

Number & Algebra

AC9M4N08

Word Problems

Complete the number sentences and write the answers for these word problems.

① Kai bought 10 mini-flags to decorate the plates of sandwiches.
The mini-flags cost 65c each. How much did Kai spend?

____ × ____ = $__·____

② The dance teacher needs three $\frac{1}{4}$ metre lengths of elastic.
How many centimetres of elastic does she need?

____ × ____ = _____ cm

③ It is going to cost a lot to put the show on.
Top hats cost $25 and the cast need 9 of them. How much will that cost?

____ × ____ = $______

④ The show tickets are $5·25 each. 100 tickets have been sold.
How much money is that?

_______ × _______ = $______

⑤ How many $\frac{1}{3}$ metre lengths of ribbon can the dance teacher cut from 3 metres of ribbon?

____ ÷ — = ___

SCORE /5 0-2 3-4 5

Statistics & Probability

AC9M4P01

Weather Chart

The Weather Chart for the Week

Mon.	Tue.	Wed.	Thu.	Fri.	Sat.	Sun.

What does the chart tell us about the chance of rain this week?
Complete this prediction chart, using one tick for each day to show the chance of rain.

	Impossible	Unlikely	Likely	Very likely	Certain
① Mon.	✓				
② Tue.					
③ Wed.					
④ Thu.					
⑤ Fri.					
⑥ Sat.					
⑦ Sun.					

SCORE /7 0-3 4-5 6-7

Symmetry

These shapes need completing. Each one is symmetrical, so draw the mirror image for each one.

① 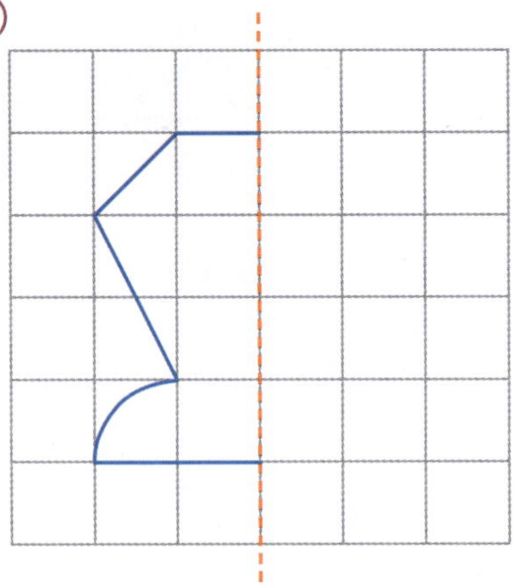

② 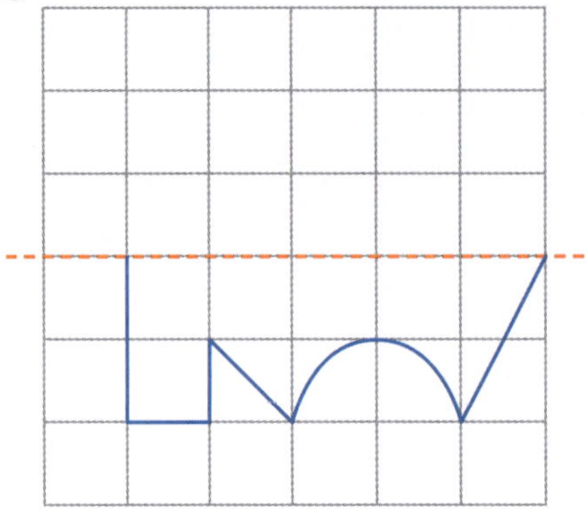

③ 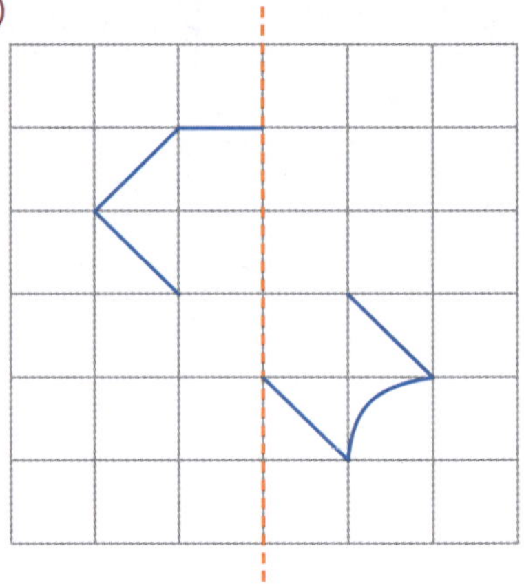

④ **Complete this design so that it has two lines of symmetry.**

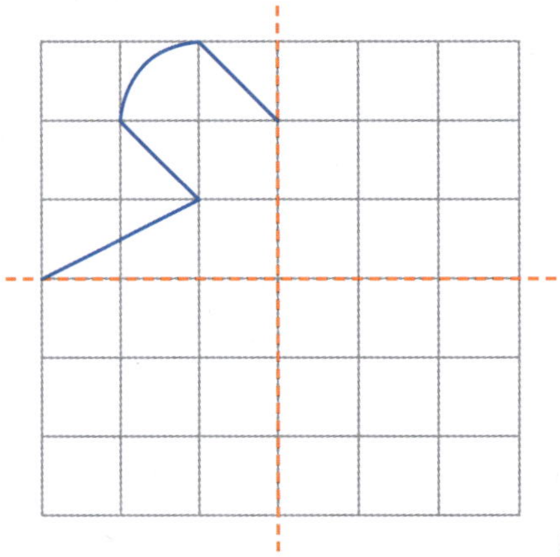

SCORE /4 0-1 2-3 4

NAPLAN*-style Questions

① Claire has made a list of the ingredients that she will need to make the fruit drink for her party.

If Claire rounds the prices to the nearest dollar, what will be her estimate of the cost?

Fruit	
Apples	$6·95
Oranges	$1·25
Lemon	95c
Juice	$3·85

② Jake has started this number pattern:
3·5, 3·7, 3·9.

What will be the next number in Jake's sequence?

③ Which of these shapes has exactly one line of symmetry?

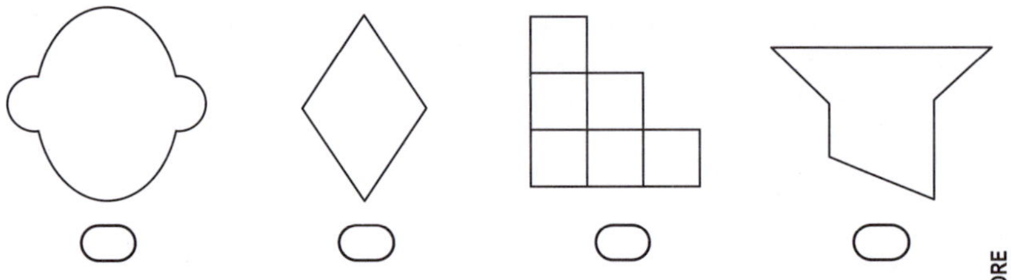

SCORE /3 0-1 2 3

TARGETING MATHS HOMEWORK: YEAR 4 © PASCAL PRESS ISBN: 9781925726572

Problem of the Week

AC9M4SP03

Symmetry with 7 Tiles

"What different symmetrical designs can you make with 7 tiles?" Mrs Green asked her class.

Claire made this design:

Chen made this design:

"Can't be done!" they said.

"Keep trying!" said Mrs Green. "I think there are lots of possibilities."

Experiment on the grid below to see if you can show Claire and Chen some designs that really do have a line of symmetry.

Here's an idea. If you can find one design that is symmetrical, try moving two of the tiles in a symmetrical way and you might be able to make a different design.

Review: Unit 9 and Unit 10

AC9M4N03, AC9M4N06, AC9M4ST01, AC9M4A02

Write the decimal that is $\frac{4}{10}$ more.

(1) 0·3 ______ (2) 0·6 ______ (3) 1·9 ______

Write the decimal that is $\frac{4}{10}$ less.

(4) $\frac{9}{10}$ ______ (5) $\frac{16}{10}$ ______ (6) $\frac{33}{10}$ ______

Continue this fraction counting sequence, using only proper fractions or mixed numbers.

(7) $1\frac{1}{4}$, $1\frac{2}{4}$, $1\frac{3}{4}$, 2, ___, ___, ___, ___

Continue this fraction counting sequence, using only proper fractions or improper fractions.

(8) $\frac{2}{3}$, $\frac{3}{3}$, $\frac{4}{3}$, ___, ___, ___, ___, ___

(9) **Use the clues to complete the graph of the Brussel Sprouts Popularity survey of 100 people.**

Five-tenths of the people hate Brussel sprouts.

20 more people don't like them than do like them.

How many people like Brussels sprouts? ______

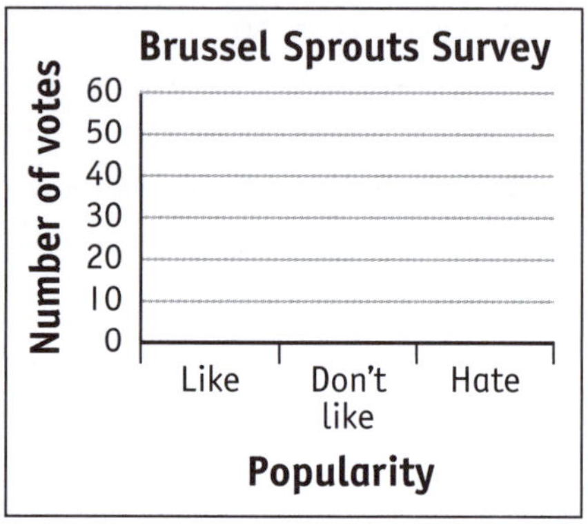

Label these diagrams in tenths, hundredths or both.

(10)
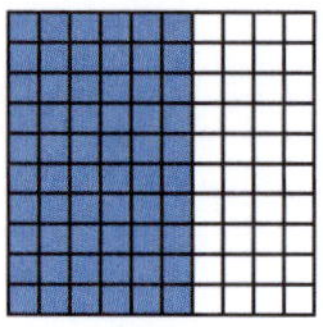
$\overline{10}$

(11)
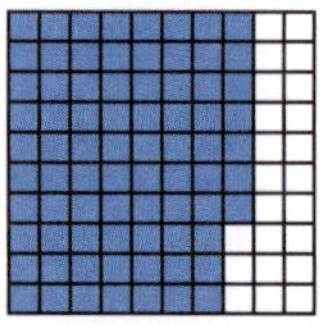
$\overline{100}$

(12)
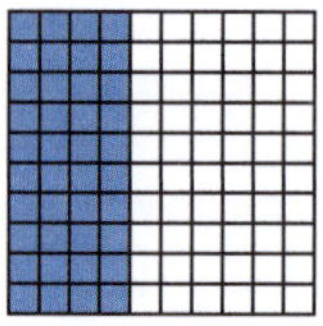
$\overline{10}$

$\overline{100}$

(13)
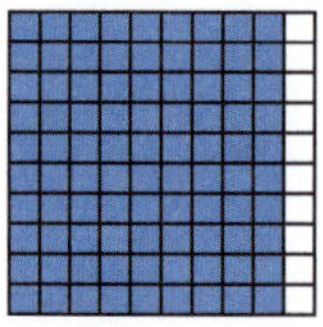
$\overline{10}$

$\overline{100}$

Complete the area models for these multiplications.

(14) 36 × 5

× ____ + ___

__ [ ] [ ] = ______

(15) 234 × 6

× ____ + ____ + ___

__ [ ] [ ] [ ] = ______

Look at the two thermometers.

(16) What is the temperature difference?

______

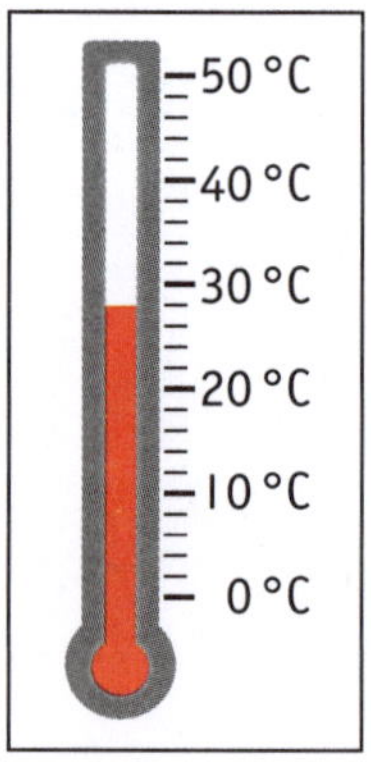

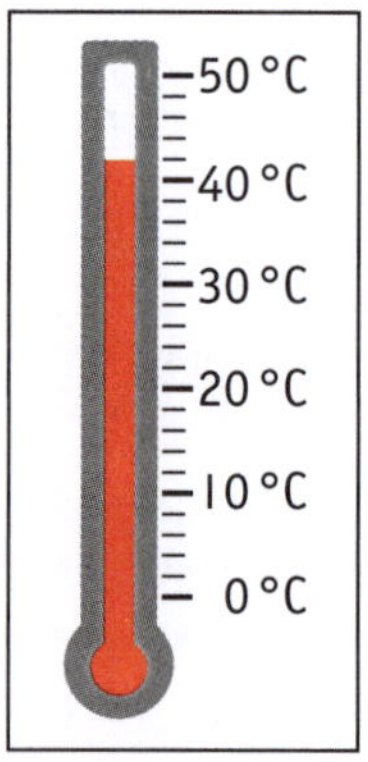

SCORE /16 0-7 8-14 15-16

TARGETING MATHS HOMEWORK: YEAR 4 © PASCAL PRESS ISBN: 9781925726572

Review: Unit 11 and Unit 12

AC9M4N01, AC9M4N05, AC9M4N06, AC9M4N08, AC9M4M01, AC9M4P01

Write these decimals as mixed numbers.

(1) 1·36 ____ (2) 1·05 ____ (3) 2·73 ____

Write these improper fractions as decimals.

(4) $\frac{18}{10}$ ___·___ (5) $\frac{112}{10}$ ___·___ (6) $\frac{208}{100}$ ___·___

Complete these number-splitting divisions.

(7) 84 ÷ 7 (Think 70 ÷ 7 = 10)

☐ + ☐

÷ 7 ↓ ↓

☐ + ☐ = ☐

(8) 72 ÷ 4 (Think 40 ÷ 4 = 10)

☐ + ☐

÷ 4 ↓ ↓

☐ + ☐ = ☐

Only one of these counters is black on the other side.

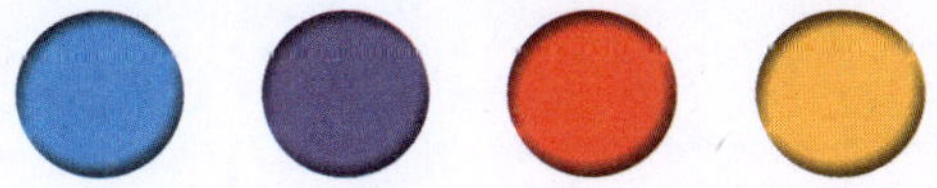

(9) How likely is it that the red counter is black on the other side?

◯ Impossible ◯ Unlikely ◯ Likely ◯ Very likely ◯ Certain

Complete these decimal conversions

(10) 1123 g = ________ kg (11) 1·5 L = ________ mL

Multiply these numbers by 10.

(12) 1·34 × 10 = ______ (13) 2·05 × 10 = ______

Multiply these numbers by 100.

(14) 1·34 × 100 = ________ (15) 2·05 × 100 = ________

Round the prices to estimate and then add.

(16) **Round to the nearest 10c.**

	Item price	Rounded price
	$1·54	$ ______
	$2·75	$ ______
	$3·69	$ ______
	$2·28	$ ______
Total:	$ ______	$ ______

(17) **Round to the nearest $1.**

	Item price	Rounded price
	$3·25	$ ______
	$4·59	$ ______
	$6·22	$ ______
	$7·53	$ ______
Total:	$ ______	$ ______

SCORE /17 (0-8) (9-15) (16-17)

TARGETING MATHS HOMEWORK: YEAR 4 © PASCAL PRESS ISBN: 9781925726572

Review: Unit 13 and Unit 14

AC9M4N01, AC9M4N02, AC9M4N06, AC9M4N08, AC9M4N09, AC9M4A02, AC9M4M01, AC9M4P01

Write these money amounts in dollar-and-cent notation.

(1) 165 cents = $ ___.___ (2) 305 cents = $ ___.___ (3) 220 cents = $ ___.___

Complete the related division facts for these multiplications.

(4) Use 5 × 3 = ☐ for ☐ ÷ 5 = ☐ and ☐ ÷ 3 = ☐

(5) Use 6 × 8 = ☐ for ☐ ÷ 6 = ☐ and ☐ ÷ 8 = ☐

(6) Use 9 × 7 = ☐ for ☐ ÷ 9 = ☐ and ☐ ÷ 7 = ☐

Write these measurements in decimal form.

(7) Two metres and sixty-four centimetres = ___._____ m

(8) Three metres and eight centimetres = ___._____ m

Will these answers be odd or even?

(9) 2 + 6 + 5 _______

(10) 4 + 14 + 28 _______

(11) 13 + 27 + 64 _______

(12) 7 × 3 _______

(13) 9 × 55 _______

(14) 8 × 65 _______

Complete these equivalent additions.

(15) 17 + 3 = 7 + ____

(16) 29 + 6 = 30 + ____

(17) 5 + 17 = 17 + ____

(18) 6 + 27 = 16 + ____

The inter-school final is scheduled for next Friday.

(19) If it storms on that day, how likely is it that the game will go ahead?

◯ Impossible ◯ Unlikely ◯ Likely ◯ Very likely ◯ Certain

(20) If it is hot and sunny on that day, how likely is it that the game will go ahead?

◯ Impossible ◯ Unlikely ◯ Likely ◯ Very likely ◯ Certain

Jo tossed a coin 10 times and recorded the results.

Heads	𝍸
Tails	𝍸

(21) What is the chance that Jo's next toss will be tails?

◯ Impossible ◯ Unlikely ◯ Likely ◯ Very likely ◯ Certain

(22) What is the chance that Jo's next toss will be heads or tails?

◯ Impossible ◯ Unlikely ◯ Likely ◯ Very likely ◯ Certain

SCORE /22 0-10 11-20 21-22

TARGETING MATHS HOMEWORK: YEAR 4 © PASCAL PRESS ISBN: 9781925726572

TERM 2 REVIEW

Review: Unit 15 and Unit 16

AC9M4N01, AC9M4N03, AC9M4N08, AC9M4A02, AC9M4SP01, AC9M4SP03, AC9M4ST03

Write the equivalent landmark decimals.

① $\frac{3}{4}$ of a metre = ______ m

② $1\frac{1}{4}$ metres = ______ m

Use the × 9 strategy for these multiplications.

③ 7 × 9 = ____ × 10 − ____ = ____

④ 14 × 9 = ____ × ____ − ____ = ____

This chart shows daily midday temperatures for last week.

⑤ Which was the hottest day?

____

⑥ Which was the coldest day?

____

⑦ How much hotter was it on Saturday than on Wednesday?

____

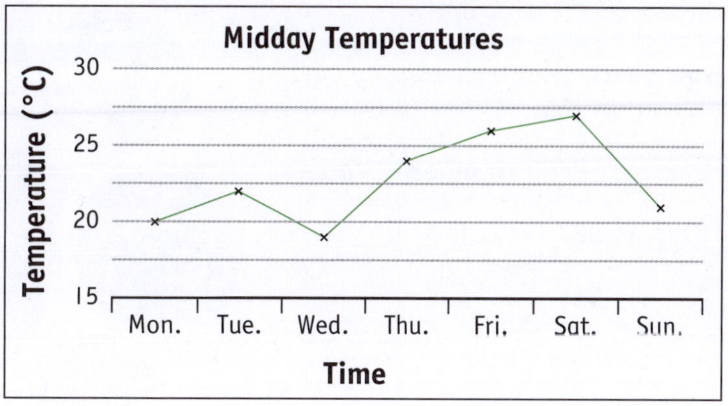

What is the area of each of these shapes?

⑧

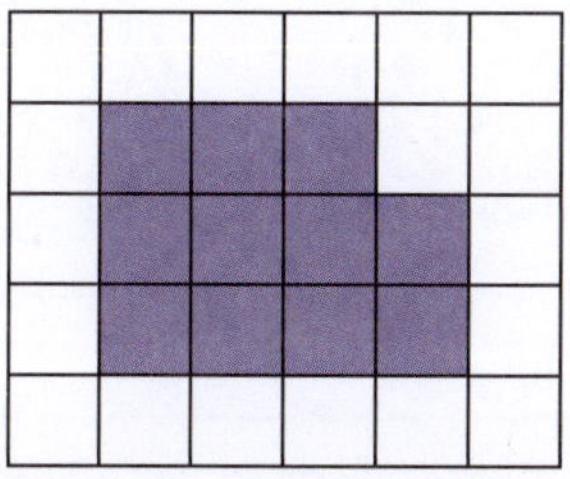

_____ squares

⑨

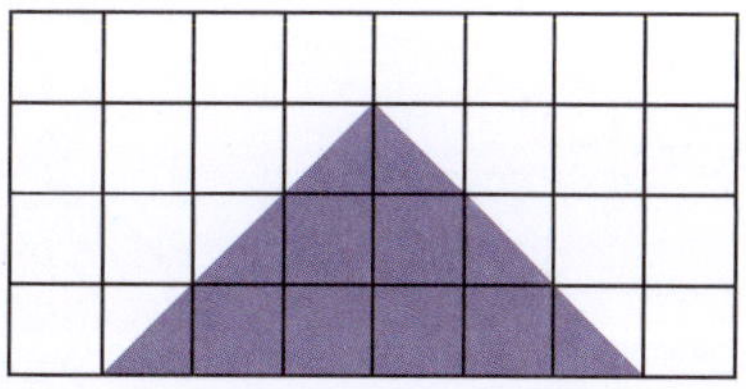

_____ squares

Complete these decimal counting sequences.

⑩

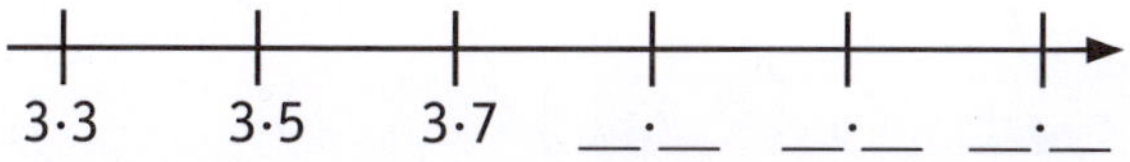

⑪

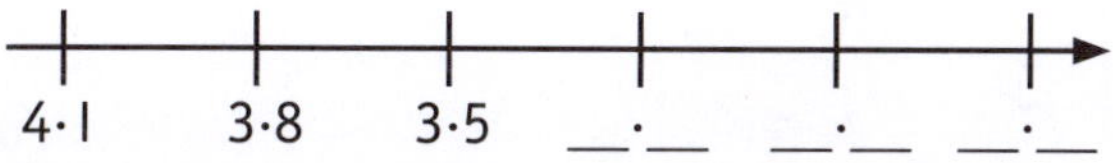

Complete the number sentences and write the answers for these word problems.

⑫ We bought 6 packets of bread rolls with 4 in each packet. How many bread rolls was that?

____ × ____ = ____

⑬ $\frac{1}{4}$ of a metre of elastic and $\frac{1}{2}$ a metre of elastic is how long altogether?

___ + ___ = ___

Complete these shapes to make them symmetrical about the orange dashed line.

⑭

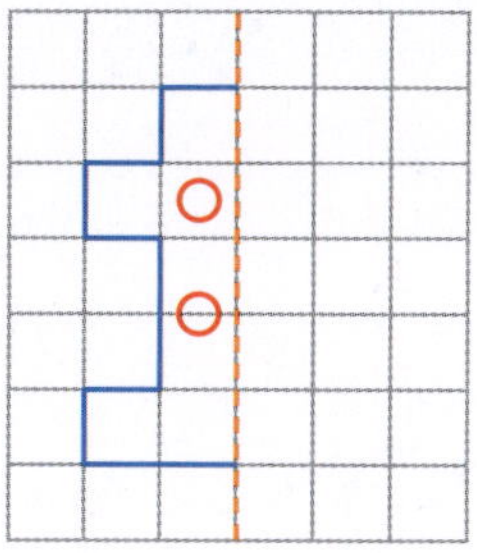

⑮

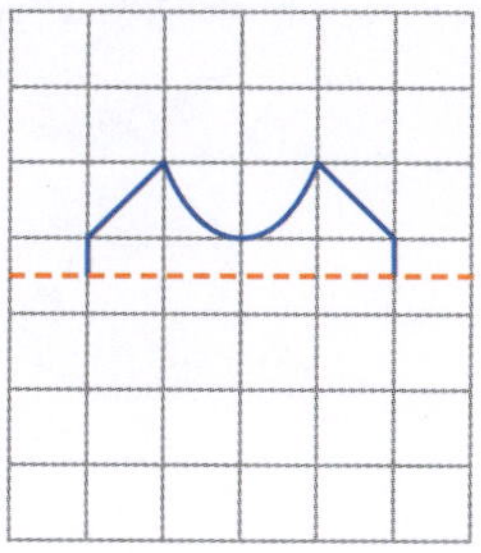

SCORE /15 0-7 8-13 14-15

TERM 2

AC9M4N01, AC9M4N07

Estimating Money Amounts

① Look at the spilt coins and estimate their total value.

- ◯ less than $5
- ◯ less than $10
- ◯ more than $10

Now check your estimate by sorting and counting.

		Number	Value
②	$2	1	$2
③	$1	2	
④	50c		
⑤	20c		
⑥	10c		
⑦	5c		
⑧	**Total**		

SCORE /8 0-3 4-6 7-8

TERM 3

Quick Check

Write the value.

① six  = ______

② five  = ______

③ eight  = ______

④ fifteen  = ______

⑤ twenty-one  = ______

⑥ 12  and 2  = ______________

⑦ 5  and 3  = ______________

⑧ 12  and 4  = ______________

⑨ 15  and 4  = ______________

⑩ 18  and 6  = ______________

SCORE /10 0-4 5-8 9-10

TARGETING MATHS HOMEWORK: YEAR 4 © PASCAL PRESS ISBN: 9781925726572

Number & Algebra

AC9M4N04

Fractions – Proper, Improper and Mixed Numbers

Write these fractions in order from smallest to largest.

1. $\frac{13}{8}$ $\frac{5}{4}$ $\frac{7}{8}$ $1\frac{1}{8}$ $\frac{3}{4}$ ____________________
2. $\frac{1}{6}$ $\frac{7}{6}$ $\frac{4}{3}$ $\frac{3}{6}$ $\frac{1}{3}$ ____________________
3. $1\frac{1}{5}$ $\frac{1}{2}$ $\frac{3}{5}$ $\frac{11}{10}$ $\frac{8}{5}$ ____________________
4. $\frac{1}{4}$ $\frac{1}{10}$ $\frac{1}{5}$ $\frac{1}{6}$ $\frac{1}{8}$ ____________________
5. $1\frac{1}{2}$ $1\frac{1}{4}$ $1\frac{1}{5}$ $\frac{3}{4}$ $\frac{3}{5}$ ____________________

SCORE /5 0-2 3-4 5

Statistics & Probability

AC9M4ST03

Votes for an Excursion

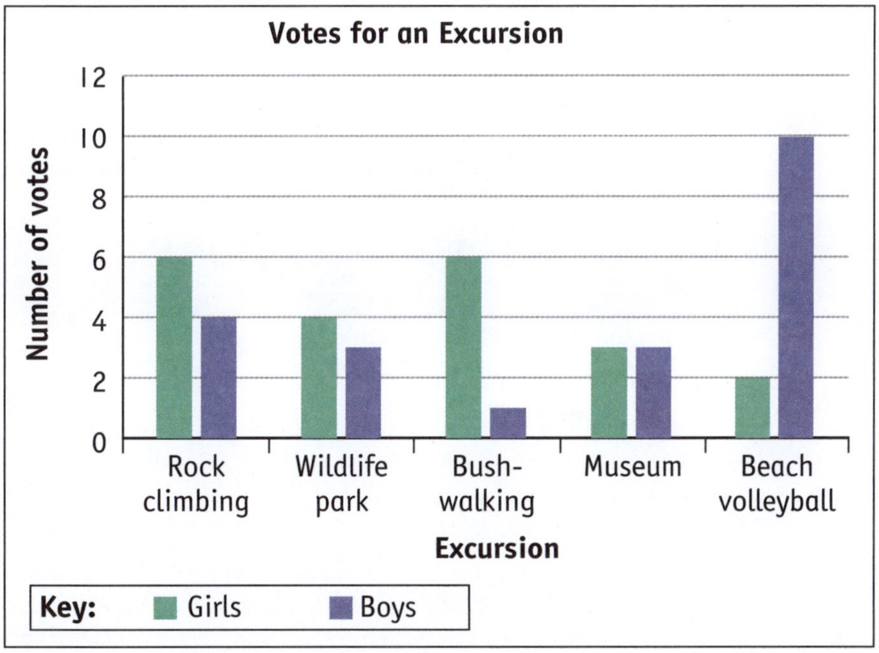

1. Were there more boys than girls in the survey? ________
2. Which was the most popular excursion? ____________________
3. Which was the least popular excursion? ____________________
4. How many students voted for bushwalking? ________
5. Do you think that volleyball would be the best choice for the excursion? ________
   Why or why not? ______________________________

SCORE /5 0-2 3-4 5

TERM 3

AC9M4M03

Time Durations

My pilates class starts at 11:35 am and finishes at a quarter to one.
To show how long the class lasts, you can use a timeline like this.

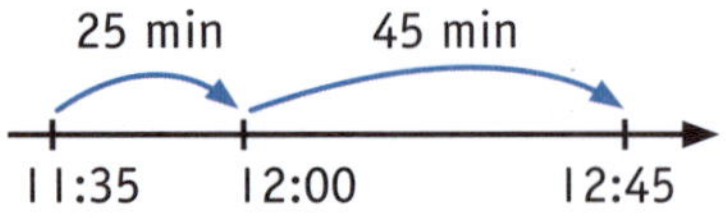

25 min + 45 min = 1 h 10 min

Notice how working out the amount of time from 11:35 to 12:00 makes it easy to find out how long the class lasts altogether.

Use the timelines to answer these questions. Answer in hours and minutes.

1. How long is a movie that starts at 6:50 pm and ends at 9:10 pm?

Total time = ____________________

2. The train journey will start at 10:20 am and get to Grandma's town at 1:15 pm. How long will the journey to Grandma's take?

Total time = ____________________

3. The cake needs 45 minutes to cook. If it goes into the oven at 11:40 am, at what time will it be cooked?

Total time = ____________________

4. Football training starts at 4:18 pm and finishes at 5:15 pm. How long is that?

Total time = ____________________

5. Jo's school day starts at 8:54 am and finishes at 3:20 pm. How long is Jo's school day?

Total time = ____________________

TERM 3

SCORE /5 0-2 3-4 5

NAPLAN*-style Questions

1. Jake has fifteen ten-cent coins and twenty-five five-cent coins. How much money does he have altogether? ☐

2. What is the missing value on this number line? ☐

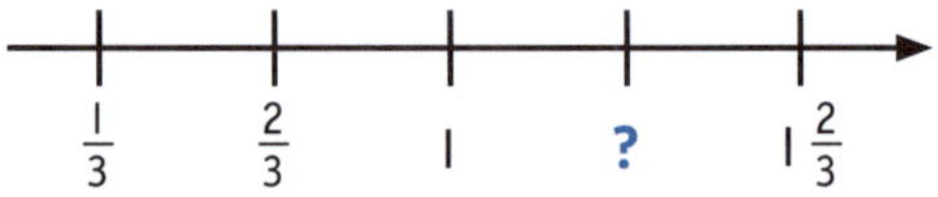

3. The movie starts at 6:55 pm and ends at 8:20 pm. How long is the movie, in minutes?

☐ minutes

SCORE /3 0-1 2 3

TARGETING MATHS HOMEWORK: YEAR 4 © PASCAL PRESS ISBN: 9781925726572

Checking their Savings

Claire, Jo and Jake each tipped out their savings which together added to $21.

Claire had:
- 18 × 5c
- 12 × 10c
- 22 × 20c
- 3 × 50c.

Jo had enough to make the total $21.
She had no gold coins and the smallest number of silver coins possible.

Jake had:
- 23 × 5c
- 15 × 10c
- 18 × 20c
- 6 × 50c.

Question Time!

1. How much money did Jo have?

   Jo had $________.

2. What coins did Jo have?

   Jo had:

   ____ × 50c

   ____ × 20c

   ____ × 10c

   ____ × 5c.

3. How much more money did Jake have than Jo?

   Jake had $________ more than Jo.

4. How much less money than Claire did Jo have?

   Jo had $________ less than Claire

Jo said that Claire and Jake should give her some money so that they all had the same amount. Jake and Claire agreed.

5. How much money should each person have now?

   ____

6. What coins should Claire give to Jo?

   Claire should give Jo:

   ____ × 50c

   ____ × 20c

   ____ × 10c

   ____ × 5c.

7. What coins should Jake give to Jo?

   Jake should give Jo:

   ____ × 50c

   ____ × 20c

   ____ × 10c

   ____ × 5c.

AC9M4N05, AC9M4A02

Extending Multiplication Facts

If you know 9 × 4 = 36 then you can easily find 9 × 40.
40 is 10 times larger than 4,
so 9 × 40 is 10 times larger than 9 × 4.

For 9 × 40,
use 9 × 4 = 36
then 9 × 4 × 10 = 360.

Complete these multiplications.

1. 8 × 50
   8 × 5 = ____
   8 × 5 × 10 = ____
2. 6 × 70
   6 × 7 = ____
   ____ × ____ × ____ = ____
3. 7 × 70
   ____ × ____ = ____
   ____ × ____ × ____ = ____
4. 9 × 90
   ____ × ____ = ____
   ____ × ____ × ____ = ____
5. 4 × 80
   ____ × ____ = ____
   ____ × ____ × ____ = ____
6. 8 × 70
   ____ × ____ = ____
   ____ × ____ × ____ = ____
7. 7 × 90
   ____ × ____ = ____
   ____ × ____ × ____ = ____
8. 7 × 80
   ____ × ____ = ____
   ____ × ____ × ____ = ____
9. 3 × 120
   ____ × ____ = ____
   ____ × ____ × ____ = ____
10. 4 × 110
   ____ × ____ = ____
   ____ × ____ × ____ = ____

SCORE / 10 (0-4) (5-8) (9-10)

Quick Check

Multiply these numbers by 10.

1. 84 ____
2. 76 ____
3. 98 ____
4. 102 ____
5. 132 ____

Answer these multiplication questions.

6. 4 × 60 = ________
7. 7 × 30 = ________
8. 8 × 40 = ________
9. 5 × 50 = ________
10. 9 × 70 = ________

SCORE / 10 (0-4) (5-8) (9-10)

TERM 3

TARGETING MATHS HOMEWORK: YEAR 4 © PASCAL PRESS ISBN: 9781925726572

Number & Algebra

AC9M4N06

A Strategy for Multiplying

If you can't remember a multiplication fact, you can use a strategy for finding it quickly.
Suppose you can't remember what 8 × 9 is.

I would round the 9 up to 10.

8 × 10 = 80, but that is 8 too much.

80 − 8 = **72**

I would use doubles.

8 × 9 = 4 × 18
= 2 × 36
= **72**

I would use tallies like this.

9 is 𝍸 ||||

8 × 𝍸 = 40
8 × |||| = 32
→ **72**

For each multiplication, show two strategies that you could use if you don't just know the multiplication fact. You get one mark for each strategy.

	Multiplication	Strategy 1	Strategy 2
1	8 × 7		
2	5 × 9		
3	8 × 14		
4	9 × 7		
5	12 × 14		

SCORE /10 0-5 6-8 9-10

Statistics & Probability

AC9M4ST03

Top Holiday Destinations

Half the people surveyed chose the Gold Coast.
Twelve people chose Cairns.
Three more chose Brisbane than chose Cairns.
Only three people chose Mt Isa.

1. Use the survey data to complete the graph.
2. How many people chose Brisbane? ______
3. How many people chose the Gold Coast? ______
4. How many more people chose Brisbane than Mt Isa? ______
5. How many people were surveyed altogether? ______

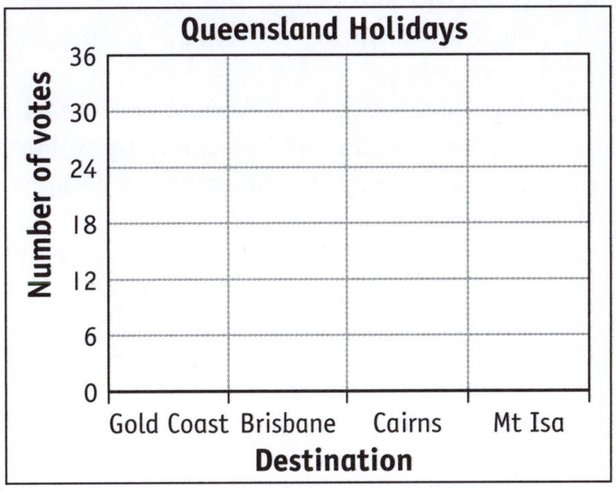

SCORE /5 0-2 3-4 5

TARGETING MATHS HOMEWORK: YEAR 4 © PASCAL PRESS ISBN: 9781925726572

AC9M4M02

Area Calculations

Each of the squares used in this shape is 1 square centimetre. These units can be written as cm^2 (centimetres squared).

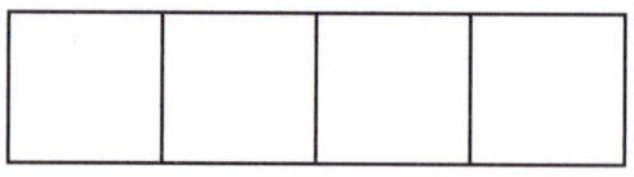

This shape has an area of 4 cm^2.

Write the area in square centimetres.

1. Area = ______ cm^2

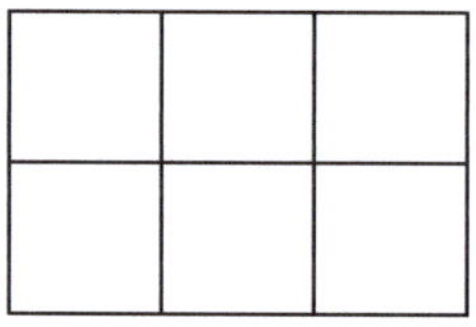

2. Area = __________

3. Area = __________

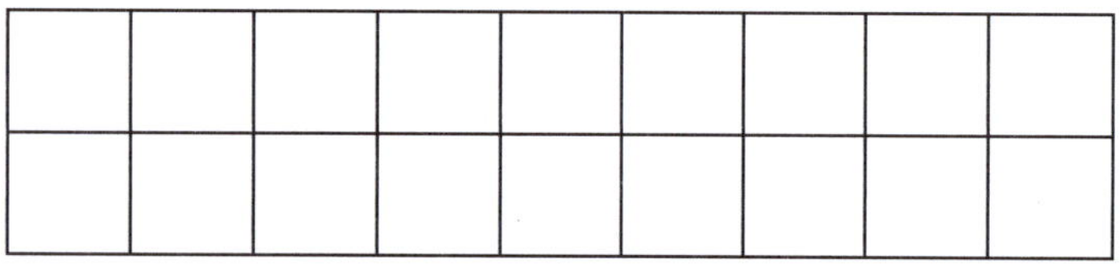

4. Area = __________

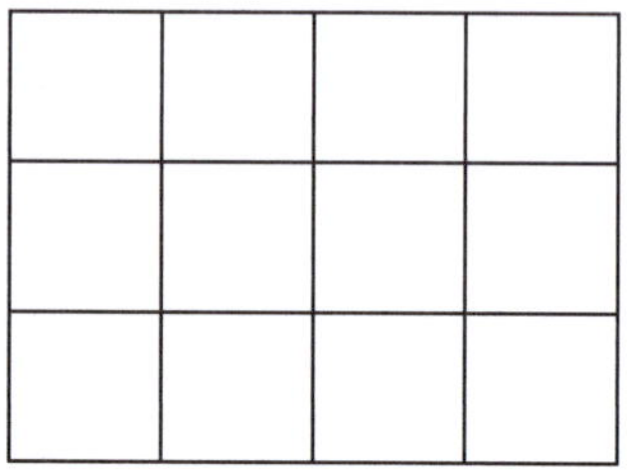

5. Area = __________

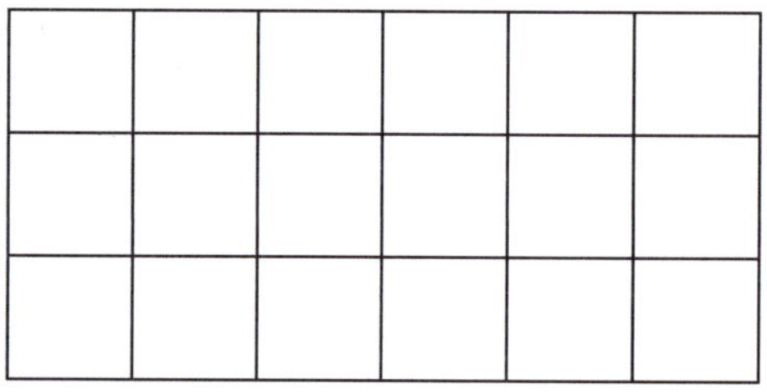

SCORE /5 (0-2) (3-4) (5)

NAPLAN*-style Questions

1. There are 240 people coming to the school concert this evening. The chairs will be set out in 12 rows. How many chairs will be in each row? ☐

2. Jake has saved $180 towards his new bike. He has been making $9 a day by walking his neighbour's dog. How many days has he walked the dog? ☐

3. Mrs Green wants to know which holiday movie they should all go to as a school treat. Who would it be best for her to ask?
   - ◯ Ask every student in the class.
   - ◯ Ask the cinema managers.
   - ◯ Ask the girls in the class.
   - ◯ Ask the boys in the class.

SCORE /3 (0-1) (2) (3)

TERM 3

TARGETING MATHS HOMEWORK: YEAR 4 © PASCAL PRESS ISBN: 9781925726572

Problem of the Week

AC9M4M02

Flood Damage

After the floods, my bedroom carpet was ruined.
I have chosen some new carpet tiles,
and they are $\frac{1}{2}$ metre square.

This is a floor plan of my bedroom.

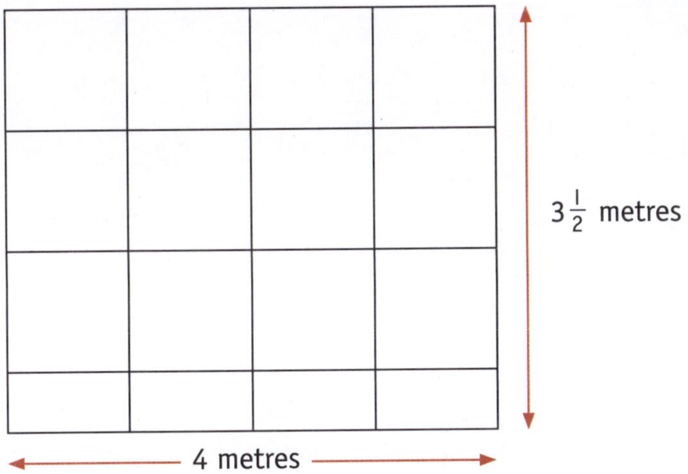

① Show on the floor plan how to work out how many carpet tiles are needed.

Chen needs ______ carpet tiles.

② Chen's budget for replacing the carpet with carpet tiles is $500.
Each carpet tile costs $7.

How much will her carpet tiles cost?
Show your working out.

$ ____________ for carpet tiles

③ Chen can spend any money left over on a new desk for her bedroom.

How much money will Chen be able to spend on a new desk?

$ _________

AC9M4N05, AC9M4A02

Extending Division Facts

If you know that **36 ÷ 9 = 4**, then you can easily find **360 ÷ 4.**
360 is 10 times larger than 36, so you can build on that.

360 ÷ 9

10 × 36 ÷ 9

10 × 4 = 40

Use Jo's method with these examples.

1. 490 ÷ 7
   10 × 49 ÷ 7
   10 × ☐ = ☐

2. 630 ÷ 9
   10 × ☐ ÷ 9
   ☐ × ☐ = ☐

3. 560 ÷ 8
   10 × ☐ ÷ 8
   ☐ × ☐ = ☐

4. 360 ÷ 6
   ☐ × ☐ ÷ ☐
   ☐ × ☐ = ☐

5. 640 ÷ 8
   ☐ × ☐ ÷ ☐
   ☐ × ☐ = ☐

6. 480 ÷ 6
   ☐ × ☐ ÷ ☐
   ☐ × ☐ = ☐

7. 720 ÷ 6
   ☐ × ☐ ÷ ☐
   ☐ × ☐ = ☐

8. 810 ÷ 9
   ☐ × ☐ ÷ ☐
   ☐ × ☐ = ☐

SCORE /8 0-3 4-6 7-8

Quick Check

Divide these numbers by 10.

1. 780 becomes ____.
2. 560 becomes ____.
3. 940 becomes ____.
4. 320 becomes ____.
5. 1040 becomes ____.

Complete these divisions.

6. 360 ÷ 4 = _____
7. 250 ÷ 5 = _____
8. 320 ÷ 8 = _____
9. 420 ÷ 6 = _____
10. 180 ÷ 9 = _____

SCORE /10 0-4 5-8 9-10

TARGETING MATHS HOMEWORK: YEAR 4 © PASCAL PRESS ISBN: 9781925726572

Number & Algebra

AC9M4A01

Finding Unknown Values in Additions

Sometimes we are given a question where one of the numbers is unknown.

- **Answer unknown:** In the addition **6 + 3 = ?**, it is the answer that is unknown.
- **Start unknown:** For **? + 3 = 9**, the starting number is unknown.
  It can be found by thinking, "What goes with 3 to make 9?"
- **Change unknown:** For **6 + ? = 9**, the number that changes 6 into 9 is unknown.
  Thinking, "What goes with 6 to make 9?" will help find the unknown number.

Find the start, change or answer unknowns in these questions.

1. ☐ + 7 = 15
2. 8 + ☐ = 12
3. 9 + 12 = ☐
4. ☐ + 26 = 49
5. 18 + ☐ = 37
6. 27 + 35 = ☐
7. ☐ + 37 = 84
8. 26 + ☐ = 90
9. 136 + 64 = ☐
10. ☐ + 53 = 107
11. 46 + ☐ = 131
12. 231 + 579 = ☐

SCORE /12 (0-5) (6-10) (11-12)

Statistics & Probability

AC9M4P02

Drop a Coin

For this fun experiment you have to drop a coin from about 10 cm over this target. I like to use a 5c coin, but you can choose your own if you like. Repeat this by dropping the coin 10 times.

1. How likely is it that the coin will land (come to rest) somewhere on the target?
   Note: If it is only half-on, that counts as out.
   ◯ Impossible ◯ Unlikely ◯ Likely
   ◯ Very likely ◯ certain
2. Run the experiment and record your results.

	Inside	Outside
1st 10 drops		
2nd 10 drops		

3. How accurate was your prediction? ______________________________
4. If you drop the coin one more time, how likely is it to come to rest somewhere on the target?
   ◯ Impossible ◯ Unlikely ◯ Likely ◯ Very likely ◯ Certain

SCORE /4 (0-2) (3) (4)

TERM 3

Lines of Symmetry

Jake, Claire and Jo have been designing logos using these four shapes:

Show any lines of symmetry in these logos.

① ____ lines of symmetry

② ____ lines of symmetry

③ ____ lines of symmetry

④ Design your own logo using those shapes.

⑤ Mark any lines of symmetry on your logo.

SCORE /5 (0-2) (3-4) (5)

NAPLAN*-style Questions

① 89 + ? = 205

What number does ? represent?

◯ 106 ◯ 115 ◯ 116 ◯ 1016

② Which of these shapes is not shown in the robot picture?

◯ Square ◯ Triangle

◯ Rectangle ◯ Circle

③ If you complete these designs by reflecting them in the line of symmetry, which completed shape is a rhombus?

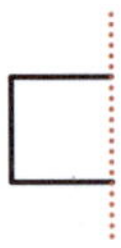
◯

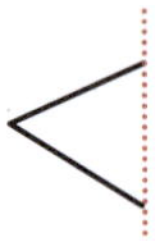
◯

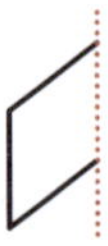
◯

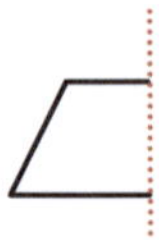
◯

SCORE /3 (0-1) (2) (3)

TARGETING MATHS HOMEWORK: YEAR 4 © PASCAL PRESS ISBN: 9781925726572

Problem of the Week

AC9M4SP03

Make a Logo Design

The brief is to make a logo design that includes one or more of each of these shapes.

The design must have at least two lines of symmetry.

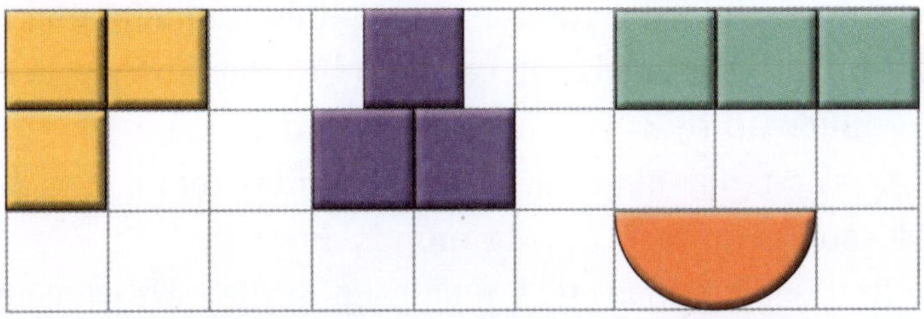

Use this grid to experiment.

Choose one that best fits the brief. Draw it in the box below and use at least two colours to make it into a symmetrically coloured logo.

Rounding Decimals

Rounding decimals is similar to rounding numbers and money.
When rounding decimals, we can round to the next whole number.
1·6 rounds up to 2. 1·3 rounds down to 1.
We can also round hundredths to the nearest tenth.
1·69 rounds up to 1·7. 1·64 rounds down to 1·6.
As with rounding numbers, 5 rounds up to the next whole or tenth.
1·5 rounds up to 2. 1·65 rounds up to 1·7.

Round these numbers to the nearest whole number.

1. 3·6 ___
2. 4·4 ___
3. 4·19 ___
4. 5·45 ___
5. 3·72 ___
6. 3·91 ___
7. 12·5 ___
8. 15·48 ___
9. 19·62 ___

Round these numbers to the nearest tenth.

10. 6·78 _____
11. 7·64 _____
12. 8·95 _____
13. 16·34 _____
14. 12·62 _____
15. 19·71 _____

Round both decimals to the nearest whole number and then add to give the estimated answer.

16. 3·4 + 6·7
Estimate = ______________
17. 3·56 + 1·73
Estimate = ______________
18. 3·5 + 7·8
Estimate = ______________
19. 4·27 + 4·83
Estimate = ______________
20. 5·2 + 3·4
Estimate = ______________
21. 16·74 + 23·68
Estimate = ______________
22. 12·42 + 11·62
Estimate = ______________
23. 5·29 + 14·72
Estimate = ______________
24. 27·44 + 23·49
Estimate = ______________

SCORE /24 0-11 12-22 23-24

TERM 3

Quick Check

Round these decimals to the nearest whole number.

1. 4·8 rounds to ____________.
2. 5·25 rounds to ____________.
3. 7·89 rounds to ____________.
4. 6·09 rounds to ____________.
5. 3·28 rounds to ____________.

Round these decimals to the nearest tenth.

6. 4·85 rounds to ____________.
7. 6·37 rounds to ____________.
8. 3·52 rounds to ____________.
9. 1·72 rounds to ____________.
10. 1·98 rounds to ____________.

SCORE /10 0-4 5-8 9-10

TARGETING MATHS HOMEWORK: YEAR 4 © PASCAL PRESS ISBN: 9781925726572

Number & Algebra

AC9M4A01

Finding Unknown Values in Subtractions

Sometimes we are given a subtraction where one of the numbers is unknown.
Answer Unknown: In the subtraction **9 – 3 = ?**, it is the answer that is unknown.
Start Unknown: For **? – 3 = 6**, the starting number is unknown.
It can be found by thinking, "What minus 3 makes 6?" or "6 + 3 makes what?"
Change Unknown: For **9 – ? = 6**, the number that changes 6 into 9 is unknown. Thinking, "What goes with 6 to make 9?" will help you find the unknown number.

Find the start, change or answer unknowns in these questions.

1. ☐ – 4 = 13
2. 27 – ☐ = 12
3. 58 – 19 = ☐
4. ☐ – 15 = 31
5. 36 – ☐ = 17
6. 83 – 27 = ☐
7. ☐ – 39 = 42
8. 58 – ☐ = 19
9. 105 – 89 = ☐

SCORE /9 0-4 5-7 8-9

Statistics & Probability

AC9M4ST02

The Class Car Colour Survey

The class found that in 30 minutes, 35 red cars, 20 white cars, 25 blue cars, 15 green cars and 40 silver cars passed the school gates.
Mrs Green told the class to show the data in a way that made it easy to interpret.
Claire chose tallies.

1. Complete each graph.

Answer these questions and circle the type of chart that you used to find the answer.

2. How many cars passed the school altogether?
___ cars, using the tallies/bar/line graph
3. How many more silver than red cars passed the school gates?
___ cars, using the tallies/bar/line graph
4. What colour were the cars in the middle of the counts?
________, ________, ________ cars,
using the tallies/bar/line graph
5. How many more blue cars would need to have passed the school gates to make blue cars as common as silver cars?
___ cars, using the tallies/bar/line graph

SCORE /5 0-2 3-4 5

Silver	
Green	
Blue	
White	
Red	

Chen chose a bar graph.

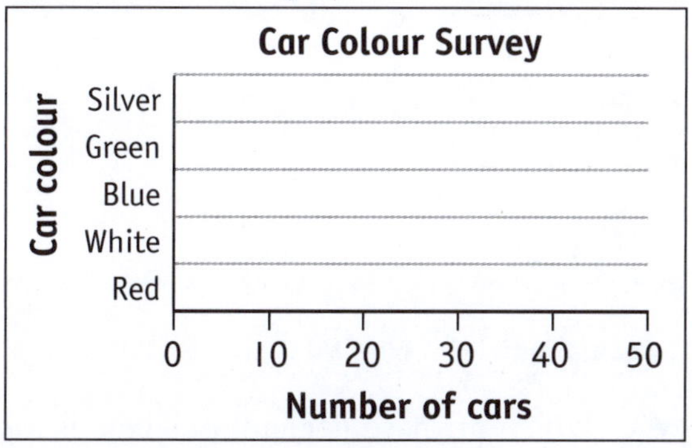

Zoe started this line graph.

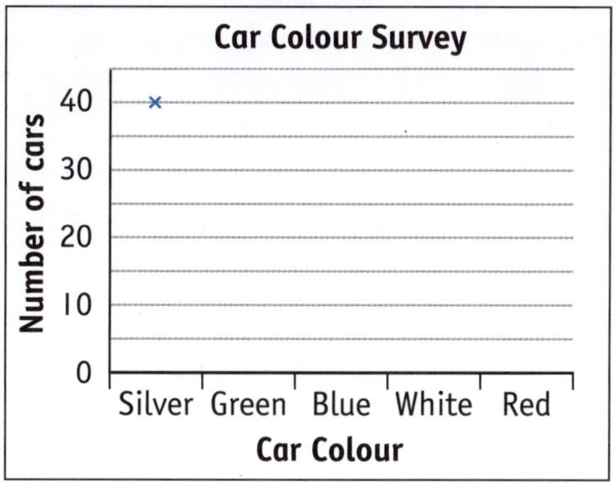

TARGETING MATHS HOMEWORK: YEAR 4 © PASCAL PRESS ISBN: 9781925726572

AC9M4M03, AC9M4M04

About Time

Write the time in words and digital form.
Then name the type of angle made by the hour and minute hands.

① 

twenty-seven past ________

____________ angle

③ 

____

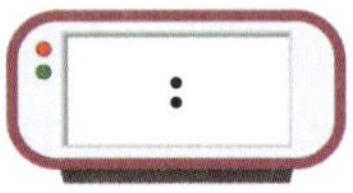

____________ angle

⑤ 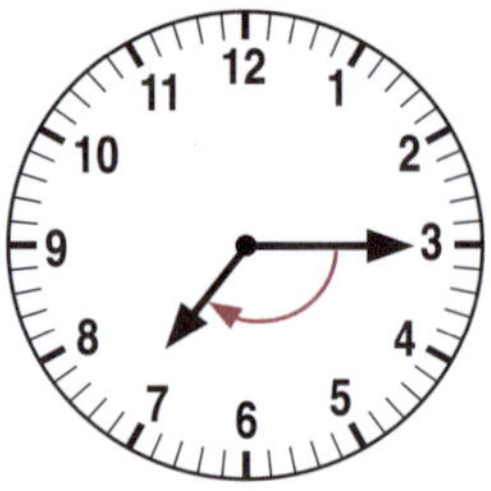

____

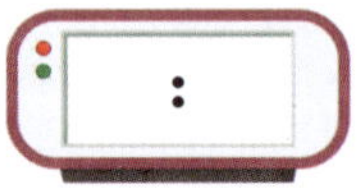

____________ angle

② 

____

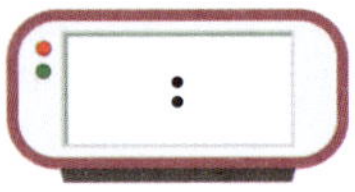

____________ angle

④ 

____

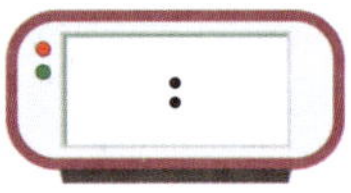

____________ angle

⑥ 

____

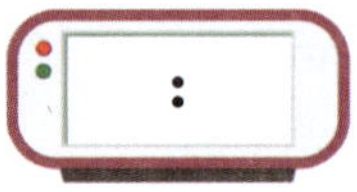

____________ angle

SCORE /6 0-2 3-4 5-6

NAPLAN*-style Questions

① Which of these decimal numbers is represented by the mark on the number line?

◯ 0·7 ◯ 0·55 ◯ 0·65 ◯ 0·75

② Which angle is about 45°?

◯ A ◯ B ◯ C ◯ D

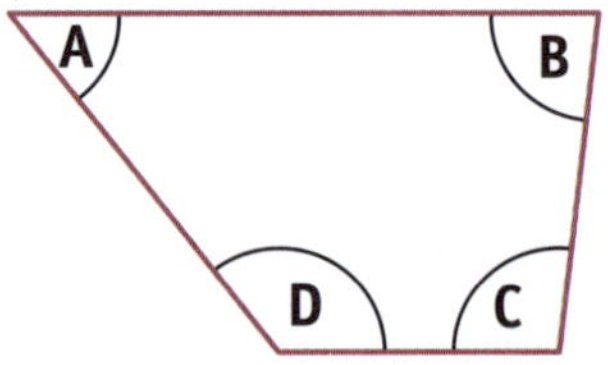

③ Complete this subtraction with the missing change unknown.

310 – ☐ = 273

SCORE /3 0-1 2 3

TARGETING MATHS HOMEWORK: YEAR 4 © PASCAL PRESS ISBN: 9781925726572

Problem of the Week

AC9M4N08

Make to 10

"Class, your task today is to make two decimal cards that add to 10.
I would like you to include tenths and hundredths on your cards.
I have made these two cards to show you what is needed.

3·65	6·35

Now it is up to you to make your own two cards.
Then you will be working in groups to match cards made by another group."

These are the cards that Claire and Jake's group made.

Match the pairs and work out what number is on the card that is covered.
Don't forget to cross off the cards as you match them.

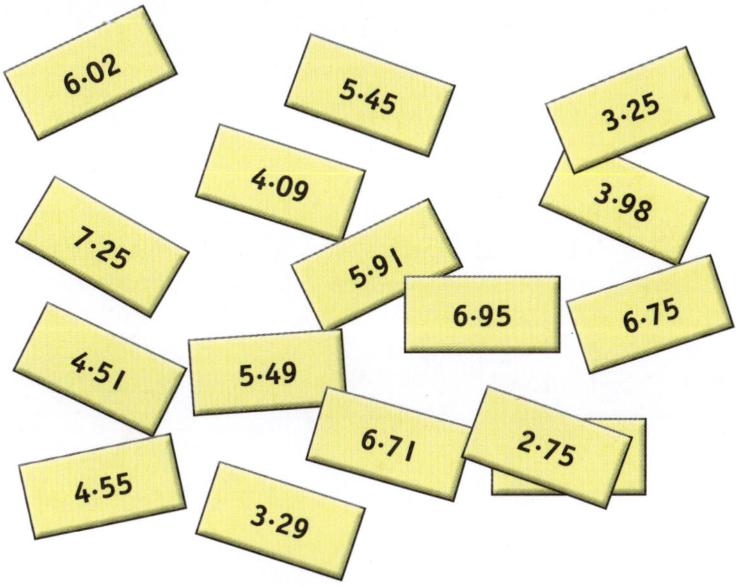

_______ + _______ = 10　　_______ + _______ = 10
_______ + _______ = 10　　_______ + _______ = 10
_______ + _______ = 10　　_______ + _______ = 10
_______ + _______ = 10　　_______ + _______ = 10

The number on the covered card is: ☐

AC9M4N03

Equivalent Fractions

One square divided into 2 equal parts	Divide each part into 2 equal parts.	Divide each part into 2 equal parts.	Equivalent fractions
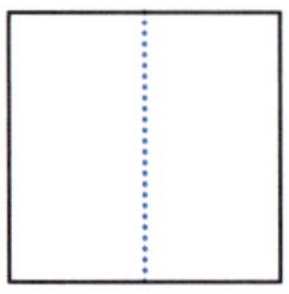	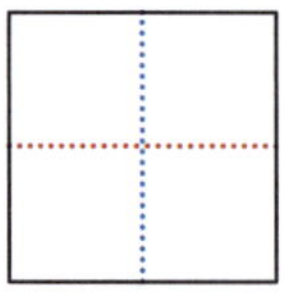	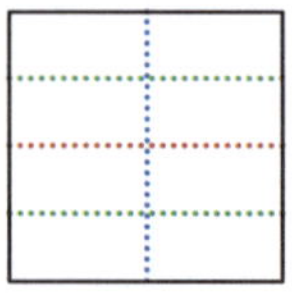	$\frac{4}{8} = \frac{2}{4} = \frac{1}{2}$ $\frac{2}{8} = \frac{1}{4}$
$\frac{2}{2} = 1$	$\frac{4}{4} = \frac{2}{2} = 1$	$\frac{8}{8} = \frac{4}{4} = \frac{2}{2} = 1$	$\frac{6}{8} = \frac{3}{4}$

Follow the steps above to complete these diagrams. Then write the equivalent fractions.

1. One square divided into 2 equal parts

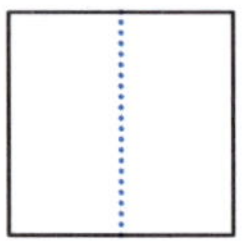

$\frac{\_\_}{2} = 1$

2. Divide each part into 3 equal parts.

$\frac{\_\_}{6} = \frac{\_\_}{2} = 1$

3. Equivalent fractions

$\frac{3}{6} = \_\_$

$\frac{2}{6} = \_\_$

4. One square divided into 3 equal parts

$\_\_ = 1$

5. Divide each part into 3 equal parts.

$\_\_ = \_\_ = 1$

6. Equivalent fractions

$\frac{3}{9} = \_\_$

$\frac{6}{9} = \_\_$

7. One square divided into 5 equal parts

$\_\_ = 1$

8. Divide each part into 2 equal parts

$\_\_ = \_\_ = 1$

9. Equivalent fractions

$\frac{2}{10} = \_\_$

$\frac{5}{10} = \_\_$

SCORE /9 0-4 5-7 8-9

Quick Check

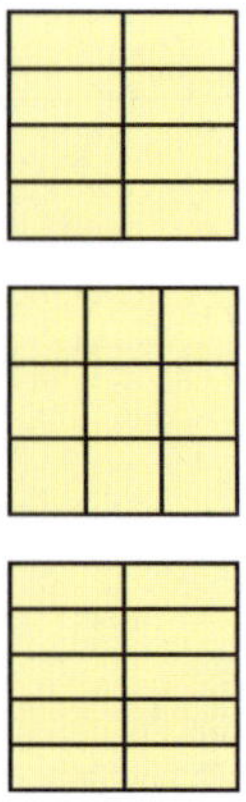

Use the diagrams to help you answer these questions.

1. $\frac{3}{4}$ is equivalent to $\frac{\_\_}{8}$
2. $\frac{2}{8}$ is equivalent to $\frac{\_\_}{4}$
3. $\frac{4}{5}$ is equivalent to $\_\_$
4. $\frac{3}{9}$ is equivalent to $\_\_$
5. $\frac{6}{9}$ is equivalent to $\_\_$
6. $\frac{5}{10}$ is equivalent to $\_\_$
7. $\frac{4}{8}$ is equivalent to $\_\_$
8. $\frac{4}{10}$ is equivalent to $\_\_$

SCORE /8 0-3 4-6 7-8

TERM 3

TARGETING MATHS HOMEWORK: YEAR 4 © PASCAL PRESS ISBN: 9781925726572

Number & Algebra

AC9M4N08, AC9M4A01

Finding Unknown Values in Money Problems

Sometimes a diagram can help sort out a problem with unknowns.

My friend Claire had $10. She bought an apple for $1·50 and a pear. She showed me that she had $7·50 left and then challenged me to find out the price of the pear. I used a diagram to find that unknown.

I added the two amounts $1·50 and $7·50 to make $9 then $10 – $9 = $1, which is how much Claire spent on the pear.

$10		
$1·50	?	$7·50
Apple	Pear	Change

Use the diagrams to solve these money problems.

① Chen spent $1·50 on an apple and now she has $3 left.

How much money did she start with?

Apple	Left over

② Jake bought a drink for $2·20 and a muffin for $4·50. He now has $6 left.

How much money did he start with?

③ Kai paid $17 to get into the movies and bought a bag of popcorn during the interval. He has $7 left out of the $30 he started with. How much was the popcorn?

④ Jo has saved $57. She will also earn $9 walking the dog and get her pocket money so that by the end of the week, she will have $86. How much is Jo's pocket money?

SCORE /4 0-1 2-3 4

Statistics & Probability

AC9M4P01

Just by Chance

I put some coloured counters into a bag and then took one out.
I recorded its colour and put it back in the bag and gave the bag a good shake!
Then I repeated the experiment.

These are my results so far:

Red	Blue	Yellow
III	III	IIII

① How many times has a counter been taken out of the bag? ______

② Which counter was picked most often? ________

③ How likely is it that the next counter will be: red? ______ blue? ______ yellow? ______

④ Explain your thinking. ____________________

SCORE /4 0-1 2-3 4

AC9M4SP02

The Unfinished Map

I have designed this map on my computer for a story that I am writing. But there are a few items missing.

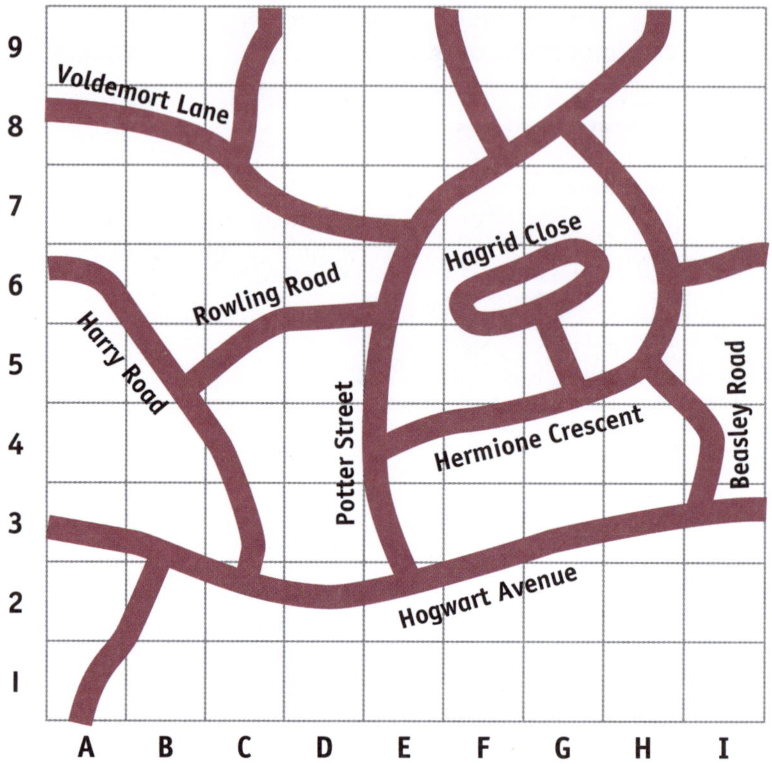

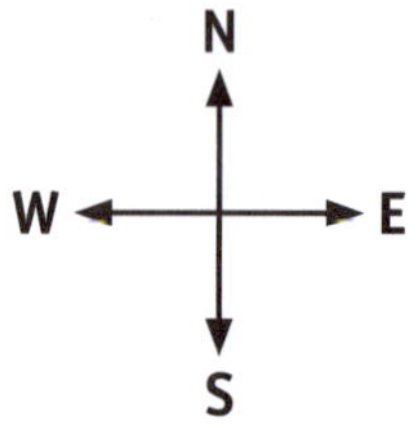

Show these items on Kai's map to finish it for him.

1. Three trees at H2
2. A cafe at D7, north of Voldemort Lane
3. A police station on the west side of Hagrid Close at F5
4. A hospital on the north side of Hogwort Avenue at G3
5. A playground on the east side of Harry Road at C3

SCORE /5 0-2 3-4 5

NAPLAN*-style Questions

1. Complete this number sentence. $12·50 – ☐ = $7·25
2. Which of these fractions is equivalent to $\frac{3}{4}$?

   ◯ $\frac{1}{3}$ ◯ $\frac{1}{2}$ ◯ $\frac{4}{6}$ ◯ $\frac{6}{8}$
3. Jo was walking north and then turned a quarter turn to her right. In which direction is Jo facing now?

   ◯ North ◯ South ◯ East ◯ West

SCORE /3 0-1 2 3

TARGETING MATHS HOMEWORK: YEAR 4 © PASCAL PRESS ISBN: 9781925726572

AC9M4N08

The Darts Tournament

- In a game of darts, the **sector** of the board shows what the dart will score.
- A dart that lands in the outer ring scores **double** the score of the sector.
- A dart that lands in the inner ring scores **treble** the score of the sector.
- A dart that lands in the middle ring scores 50 – **bullseye**!
- In a game of 301, the player throws three darts and subtracts the scores from 301. The first player to reach 0 is the winner.

After the first round of the tournament:

- Player 1 was down to 201
- Player 2 was down to 220
- Player 3 was down to 198.

When I asked the players what their darts scored in the first round, all they would tell me was that there was:

- a treble 20
- a treble 7
- a treble 16
- a 20
- a 17
- a double 16
- a double 18
- a bullseye
- a miss (the dart didn't hit the board).

Which player made which scores with their three darts?

Player 1 Scores		Player 2 Scores		Player 3 Scores

Extra Challenge: After two more rounds, each player could be on 0 after throwing 6 more darts. How could this happen?

Player 1 Scores		Player 2 Scores		Player 3 Scores

Addition by the Front-End Method

In the real world, it is important to get into the ball park with your first estimate.
This means starting with the largest numbers, which is what the **front-end method** does.

To find 146 + 339:

Start by adding the **hundreds.**	Then add the **tens.**	Then add the **ones.**	Then **add** to complete the addition.
146	146	146	146
+ 339	+ 339	+ 339	+ 339
400	400	400	400
	70	70	70
		15	+ 15
			485

Use the front-end method with these two-digit additions.

①	②	③	④
26	84	37	75
+ 39	+ 23	+ 56	+ 36
50			
+ 15	____	____	____
____	____	____	____

Use the front-end method with these three-digit additions.

⑤	⑥	⑦	⑧
136	275	436	647
+ 223	+ 332	+ 175	+ 485
300			
50			
+ 9	____	____	____
____	____	____	____

SCORE /8 0-3 4-6 7-8

Quick Check

Try using the front-end method in your head for these additions.

① 26 + 33 = ______
② 34 + 53 = ______
③ 72 + 35 = ______
④ 67 + 61 = ______
⑤ 45 + 54 = ______
⑥ 27 + 34 = ______
⑦ 18 + 33 = ______
⑧ 28 + 16 = ______
⑨ 102 + 45 = ______
⑩ 105 + 66 = ______

SCORE /10 0-4 5-8 9-10

TARGETING MATHS HOMEWORK: YEAR 4 © PASCAL PRESS ISBN: 9781925726572

AC9M4N07

Rounding Money

When we are out shopping, we often need to keep track of what we are spending. Round these costs to estimate the shopping bill for lunch.

	Round	Estimate	Actual
① 3 loaves at $3·85 each	3 @ $4	$12	________
② 6 lettuces at $3·45 each	6 @ $3·50	________	________
③ 2 chickens at $8·20 each	________	________	________
④ 4 avocados at $1·45 each	________	________	________
⑤ 3 cucumbers at $1·95 each	________	________	________
⑥	Altogether about:	☐	

⑦ Now check the actual amounts with your calculator. ☐

⑧ Was the estimate close? ______

⑨ Was the estimate higher or lower? ____________

⑩ Why do you think that was? ____

SCORE /10 0-4 5-8 9-10

Statistics & Probability

AC9M4P01

The Dalmatian Puppies

My Dalmatian is having puppies and I am very excited as well as a bit curious about the outcome.

What is the chance of these events?

Use these words:
impossible, unlikely, likely, very likely, certain.

① There will be 20 puppies. ________________

② The puppies will all be spotty. ________________

③ The puppies will all be males. ________________

④ There will be a kitten in the litter. ________________

⑤ Each puppy will weigh close to 3 kg at birth. ________________

SCORE /5 0-2 3-4 5

TARGETING MATHS HOMEWORK: YEAR 4 © PASCAL PRESS ISBN: 9781925726572

AC9M4M02

Finding Perimeters

To find the perimeter of a square, multiply the length of one side by 4.

For this shape, the perimeter is 4 × 3 m = 12 m.

3 m

3 m

To find the perimeter of a rectangle, add the two sides and multiply by 2.

2 m

4 m

For this shape, the perimeter is 2 m + 4 m = 6 m, 6 m × 2 = 12 m.

Answer these perimeter questions.
Pay attention to the units: centimetres (cm) or metres (m).

1. How much fencing is needed for a rectangular garden with long sides of 20 m and short sides of 11·5 m?

   _______ + _______ = _______

   _______ × 2 = _______ metres

2. How much electric wire is needed to go around a chicken pen with sides of 6·5 metres?

   _______ × 4 = _______ metres

3. If you draw a red line around the edge of a piece of paper with sides that are 12 cm long, how long will the line be?

   _______ × _______ = _______ cm

4. How much tape is needed to go round a rectangular sheet of glass with sides of 75 cm and 60 cm to make it safe to carry?

   _______ + _______ = _______

   _______ × _______ = _______ metres

5. A run round a rectangular sports field that has sides of 135 m and 125 m will be how many metres long?

   _______ + _______ = _______

   _______ × _______ = _______ metres

SCORE /5 0-2 3-4 5

NAPLAN*-style Questions

1. Claire's family have saved $345 towards the holiday and Jake's family have saved $256.

   How much is in the holiday account so far? $ ______

2. This is the garden plan.

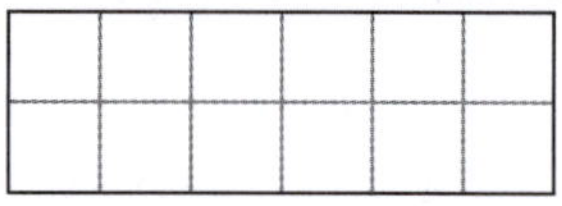

□ = 1 m^2

   What is the perimeter of the garden?

   ◯ 16 m ◯ 30 m ◯ 32 cm ◯ 8 m

3. Dan spun the spinner 8 times and these are the results: blue, red, blue, blue, red, blue, red, blue.

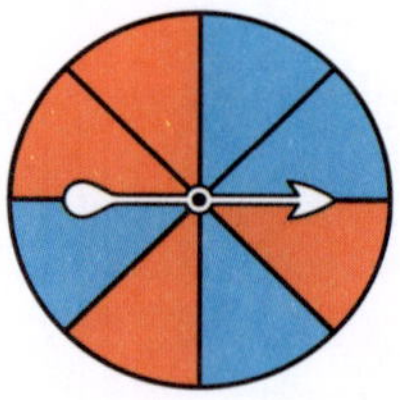

   What is most likely to happen when he spins again?

   ◯ It will be red.

   ◯ It will be blue.

   ◯ It could be either red or blue.

SCORE /3 0-1 2 3

TARGETING MATHS HOMEWORK: YEAR 4 © PASCAL PRESS ISBN: 9781925726572

Problem of the Week

AC9M4N08

A Rainy Day Game

On rainy days, Zoe, Dan and Jo like to play Drop a Coin.

It is a game they invented and it is played on a grid like this.

Players hold a coin about 5 cm above the centre of the grid and drop it. They have three turns and then add their scores.

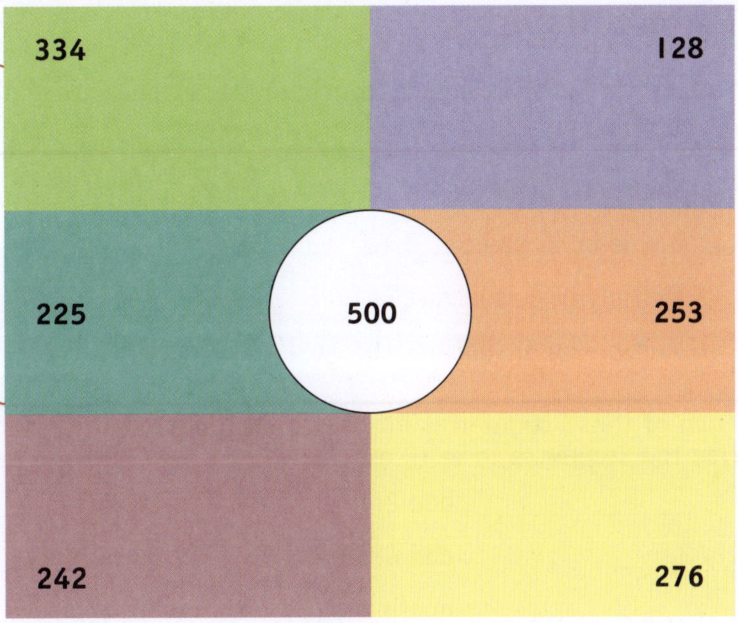

First Round

(1) Zoe got 253, 334 and 276. What was her score? Zoe's total = _____

(2) Dan scored 646. What were her three drops? Dan dropped _____, _____ and _____

(3) Jo got 225, 128 and 500. What was her score? Jo's total = _____

(4) Who was in the lead at the end of the first round? ___________

Second Round

(5) Zoe scored 629. What were her three drops? Zoe dropped _____, _____ and _____

(6) Dan scored 812. What were her three drops? Dan dropped _____, _____ and _____

(7) Jo got 225, 128 and 242. What was her score? Jo's total = _____

(8) Who is in the lead now? ___________

Your Round

Draw up a grid that is nearly the size of this page. Mark the numbers on it and play the game.

(9) After your first three drops, what was your score? _____

(10) Where would you be coming in the game with Zoe, Dan and Jo?

◯ First ◯ Second ◯ Third ◯ Fourth

(11) Play a second round of three drops. Where would you be coming now?

◯ First ◯ Second ◯ Third ◯ Fourth

TARGETING MATHS HOMEWORK: YEAR 4 © PASCAL PRESS ISBN: 9781925726572

Pattern Rules

Follow the pattern rules to continue these sequences.

1. Multiply by 2, add 3. 36, ____, ____, ____, ____
2. Subtract 3, multiply by 2. 23, ____, ____, ____, ____
3. Divide by 2, add 5. 50, ____, ____, ____, ____
4. Multiply by 5, subtract 5. 20, ____, ____, ____, ____
5. Multiply by 5, subtract 10. 60, ____, ____, ____, ____

Each of these sequences has a two-step pattern rule but you are only told part of the rule. Use the sequence to work out the rule. Then continue the sequence.

6. ____________, add 2. 28, 56, ____, ____, ____
7. ____________, add 4. 9, 36, ____, ____, ____
8. ____________, multiply by 2. 14, 24, ____, ____, ____
9. ____________, multiply by 4. 4, 13, ____, ____, ____
10. ____________, subtract 10. 5, 30, ____, ____, ____

Each of these sequences has a two-step pattern. Use the sequence to work out the pattern rule.

11. 7, 14, 19, 38, 43 Pattern rule: ____________, ____________
12. 8, 5, 10, 7, 12 Pattern rule: ____________, ____________
13. 26, 13, 12, 6, 5 Pattern rule: ____________, ____________
14. 1, 2, 6, 7, 21 Pattern rule: ____________, ____________

TERM 3

SCORE /14 0-6 7-12 13-14

Quick Check

Follow the rule to find the next two numbers.

1. Add 3, multiply by 1. 36, ____, ____
2. Subtract 3, multiply by 2. 74, ____, ____
3. Multiply by 3, subtract 2. 21, ____, ____
4. Divide by 2, add 6. 32, ____, ____
5. Divide by 3, subtract 9. 27, ____, ____

Fill in the gaps.

6. 24, multiply by 2, ____________ = 37
7. 36, ____________, add 5 = 17
8. 24, ____________, add 6 = 54
9. 27, add 3, ____________ = 3
10. 10, ____________, add 11 = 101

SCORE /10 0-4 5-8 9-10

TARGETING MATHS HOMEWORK: YEAR 4 © PASCAL PRESS ISBN: 9781925726572

Number & Algebra

AC9M4N01, AC9M4N03

Adding with Decimals

For an addition like 6·4 + 3·5 = ? you can use place value.

First add the ones, 6 + 3 = 9, and then add the tenths, $\frac{4}{10} + \frac{5}{10} = \frac{9}{10}$ or 0·9.

6·4 + 3·5 = 9·9

Try these examples.

1. 3·5 + 4·6
   7 $\frac{11}{10}$ or 1·1
   __·__

3. 4·7 + 3·8
   __ $\frac{}{10}$ or __·__
   __·__

5. 12·6 + 15·3
   __ __·__
   __·__

2. 6·3 + 4·9
   10 $\frac{12}{10}$ or 1·2
   __·__

4. 3·9 + 6·7
   __ __·__
   __·__

6. 14·5 + 16·8
   __ __·__
   __·__

SCORE /6 0-2 3-4 5-6

Statistics & Probability

AC9M4ST01

A Shoe Shuffle

The shoe shop owner watched 100 people walk past her shop and tallied their shoe type. She found that:

- $\frac{4}{10}$ of the passers-by had lace-up shoes
- $\frac{1}{4}$ of the passers-by had shoes with no fasteners
- $\frac{1}{5}$ of the passers-by had shoes with buckles.

The rest of the passers-by had Velcro fasteners.

The owner needed to make a graph to sort out the information.

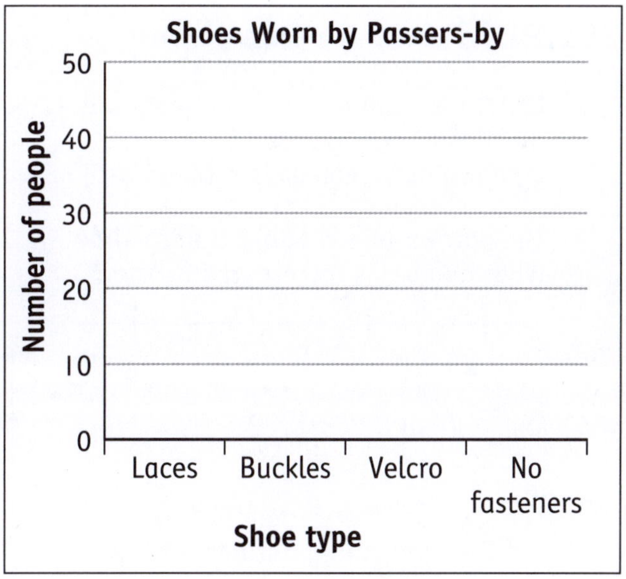

1. Complete the graph for the shoe shop owner.
2. How many people had laces? ______
3. How many people had buckles? ______
4. How many had velcro? ______
5. How many had no fasteners? ______

SCORE /5 0-2 3-4 5

AC9M4M02

Finding Areas

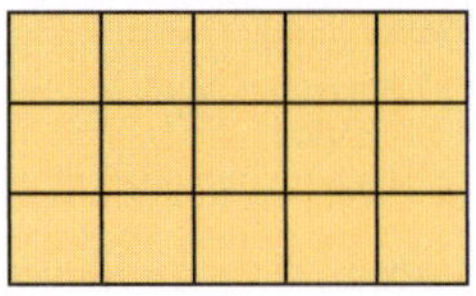

To work out the **area** of this rectangle, you could count every square. TOO SLOW!

Or you could count how many in the first row: 1, 2, 3, 4, 5. And then add 5 for the second row and 5 more for the third row. STILL TOO SLOW!

Much faster would be to say, "3 rows of 5 is 15 squares." because **3 × 5 = 15**.

Try using multiplication for these area problems.

1. 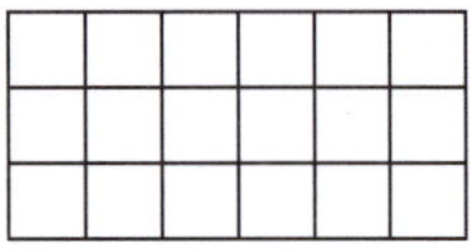

   Area = 3 × 6 = ____

2. 

   Area = ________________

3. 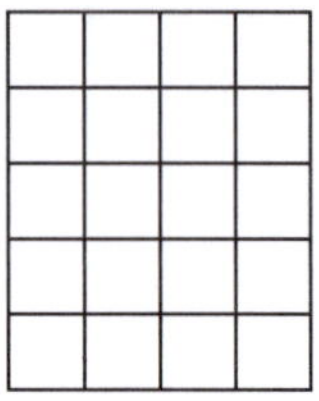

   Area = ________________

4. 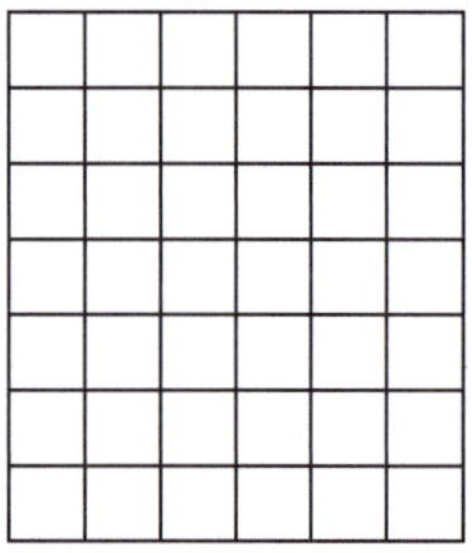

   Area = ________________

5. 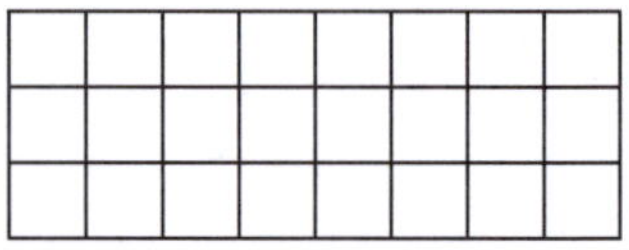

   Area = ________________

6. 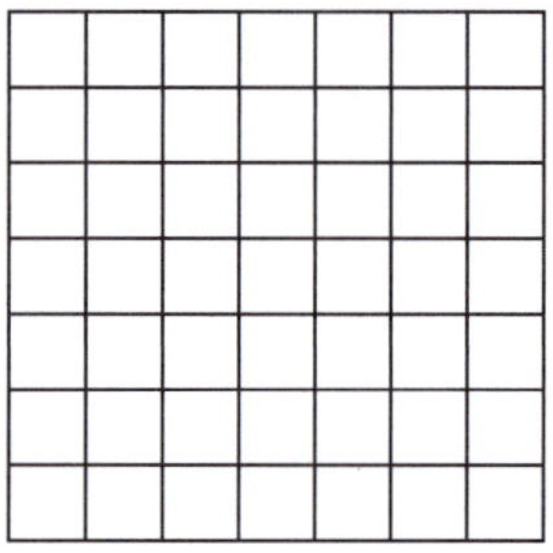

   Area = ________________

SCORE /6 0-2 3-4 5-6

TERM 3

NAPLAN*-style Questions

1. Early this morning, the classroom was 17·8 °C.
   By midday, it was 21·2 °C.
   By how many degrees had the classroom warmed up? [°C]

2. The pattern rule is **add 1·5 each time**.
   What number is represented by the **?**

   [? =]

3. For 5 weeks, the chicken farmer has been collecting data about the number of foxes seen near her chickens. Here is her graph. How many foxes were sighted in 5 weeks?

   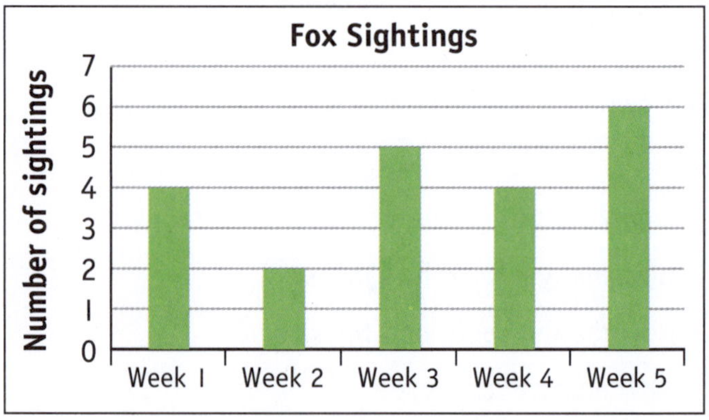

   [   ]

SCORE /3 0-1 2 3

TARGETING MATHS HOMEWORK: YEAR 4 © PASCAL PRESS ISBN: 9781925726572

Problem of the Week

AC9M4M02

Making Rectangles

"You have some grid paper here to help you make some rectangles with an area of 24 squares," Mrs Green told the class.

"But before you start, I would like you to predict whether the rectangles will:

◯ **A** all have the same perimeter

◯ **B** have perimeters of very similar length

◯ **C** have very different perimeters.

Select one prediction and then get to work making as many rectangles as possible with an area of 24 squares."

Choose A, B or C from Mrs Green's options and then make rectangles on this grid paper.

Just checking

1. How close was your prediction?

   ____

2. What is the difference between the shortest perimeter and the longest perimeter?

   ____

3. Use what you have discovered about area and perimeter to predict which rectangle of 12 squares in size will have the longest perimeter.

   ◯ 1 × 12 ◯ 2 × 6 ◯ 3 × 4

4. What is the reason for your prediction?

   ____

5. Now test it.

AC9M4N03

Fractions and Division

Each diagram has a different number sentence but they are all part of the same fact family for multiplication, division and fractions.

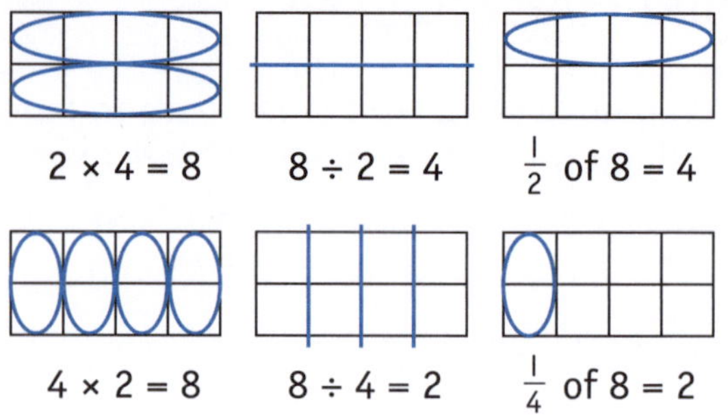

Mark and label these diagrams in the same way.

1. ___ × ___ = ___
2. ___ × ___ = ___
3. ___ ÷ ___ = ___
4. ___ ÷ ___ = ___
5. — of ___ = ___
6. — of ___ = ___

Now mark and label these diagrams in the same way.

7. ___ × ___ = ___
8. ___ × ___ = ___
9. ___ ÷ ___ = ___
10. ___ ÷ ___ = ___
11. — of ___ = ___
12. — of ___ = ___

SCORE /12 0-5 6-10 11-12

Quick Check

Write 2 division sentencess for each multiplication.

1. 6 × 3 ___ ÷ ___, ___ ÷ ___
2. 4 × 5 ___ ÷ ___, ___ ÷ ___
3. 3 × 7 ___ ÷ ___, ___ ÷ ___
4. 8 × 6 ___ ÷ ___, ___ ÷ ___
5. 9 × 5 ___ ÷ ___, ___ ÷ ___

Give one related fraction sentence for each multiplication.

6. 3 × 6 — of ___ = ___
7. 4 × 5 — of ___ = ___
8. 7 × 3 — of ___ = ___
9. 6 × 8 — of ___ = ___
10. 5 × 9 — of ___ = ___

SCORE /10 0-4 5-8 9-10

TERM 3

TARGETING MATHS HOMEWORK: YEAR 4 © PASCAL PRESS ISBN: 9781925726572

Number & Algebra

AC9M4N06

Finding the Difference

Finding the difference between two numbers is just like finding the space between them on a number line.

The number lines I've drawn show two ways of finding the space between 18 and 37.

You can use **addition** and count up from 18 to 37.

Or you can use **subtraction** and count back from 37 to 18.

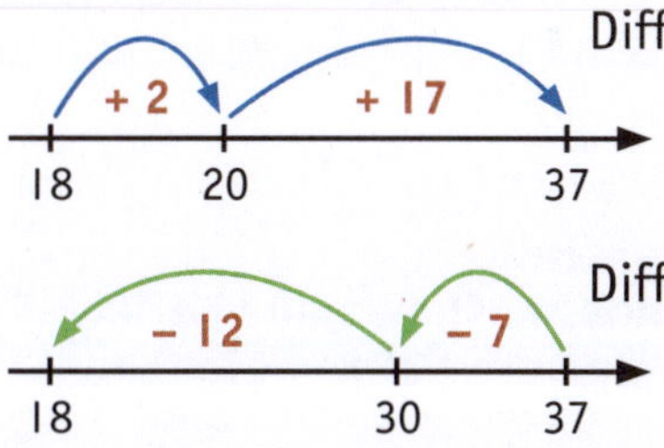

Use both addition and subtraction on the number line for each of these examples.

1. The difference between 26 and 54 is _____.
2. The difference between 23 and 41 is _____.
3. The difference between 99 and 127 is _____.
4. The difference between 72 and 101 is _____.
5. The difference between 58 and 160 is _____.
6. The difference between 67 and 90 is _____.

SCORE /6 0-2 3-4 5-6

Statistics & Probability

AC9M4ST02

Training for the Triathlon

The triathlon competitors are in training at the pool. It is early days in the training period and the competitors can't swim 50 laps yet.

This dot plot shows the number of laps the competitors did swim yesterday.

Number of Laps Swum

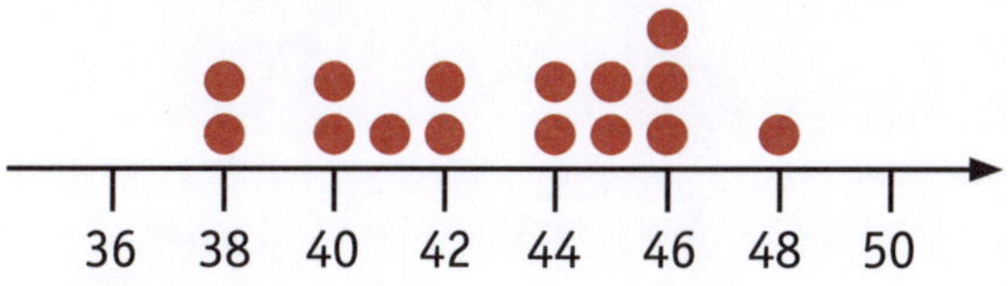

1. How many competitors are in the triathlon? _____
2. Which number of laps did most competitors swim? _____
3. How many competitors swam 43 laps? _____
4. How many competitors swam more than 45 laps? _____
5. How many competitors swam less than 45 laps? _____

SCORE /5 0-2 3-4 5

AC9M4M01

Calculating Mass

Here is an example which shows several apples on a set of scales.
The apples have the mass shown on the dial of the scales.
Use this information to help you with the following questions.

Mass: 1 kg

Read each scale carefully.
Then write what mass is shown on the scale
and how many apples are being weighed.

① ___ apples

Mass: ________

② ___ apples

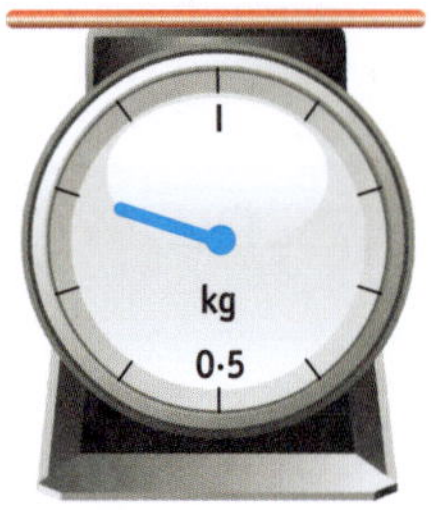

Mass: ________

③ ___ apples

Mass: ________

④ ___ apples

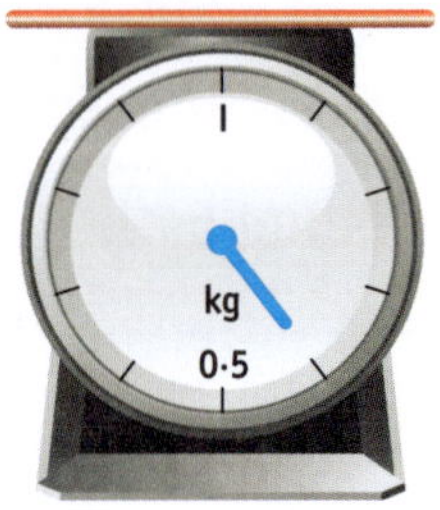

Mass: ________

SCORE /4 0-2 3 4

NAPLAN*-style Questions

① There are 40 coloured pencils in the class pencil box.
One-quarter of them are blue. How many blue pencils are there?

⬭ 4 ⬭ 40 ⬭ 10 ⬭ 30

② Kai was comparing kitchen objects on a pan balance.

Which object was the heaviest?

⬭ 
⬭ 
⬭ 
⬭ 

③ This is a dot plot of our class spelling scores.

How many students scored less than 24?

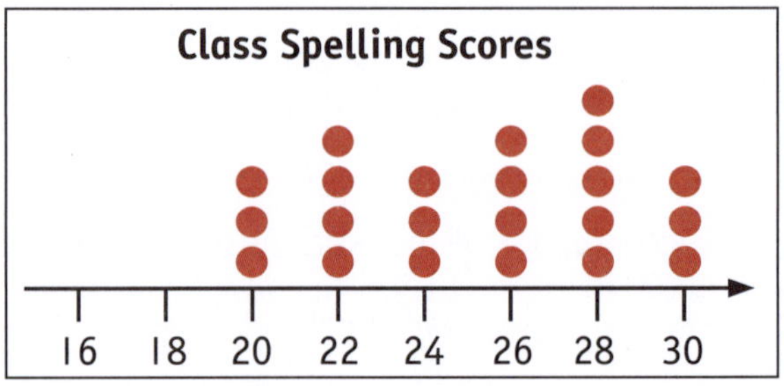

SCORE /3 0-1 2 3

TARGETING MATHS HOMEWORK: YEAR 4 © PASCAL PRESS ISBN: 9781925726572

Difference Squares

We have invented this difference puzzle.

We have tried lots of different starting numbers and we think we have found a pattern.

Here are the steps:

- Write any four numbers in the corner circles of the outside square.
- At the centre of each edge, write the difference between the two numbers that are in the corner circles.
- Repeat Step 2 for the next square, and the next square, and the next square, and the next square.

Time to try the Difference Squares puzzle

(1) It's time for you to have a go.
We suggest you start by choosing from the numbers 1–9 for the corners of the outside square.

First Try

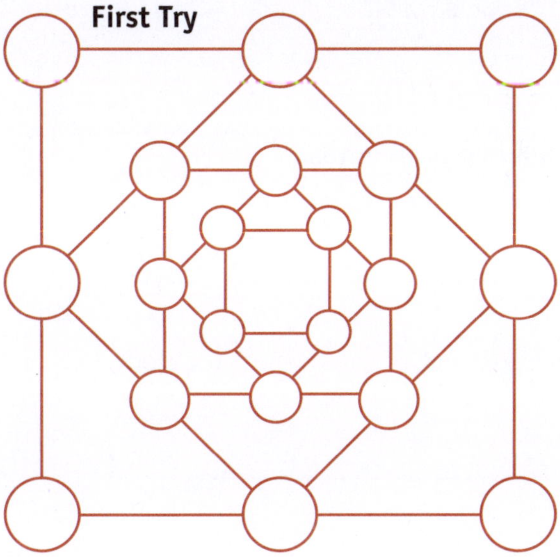

Second Try

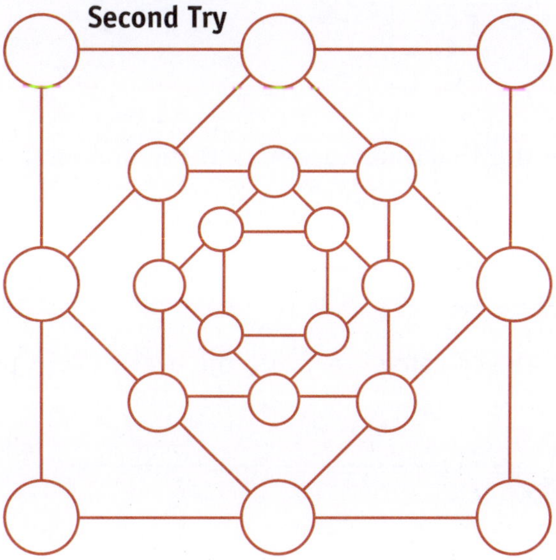

(2) Choose some larger numbers to start with. Numbers up to 20 would be good.

Third Try

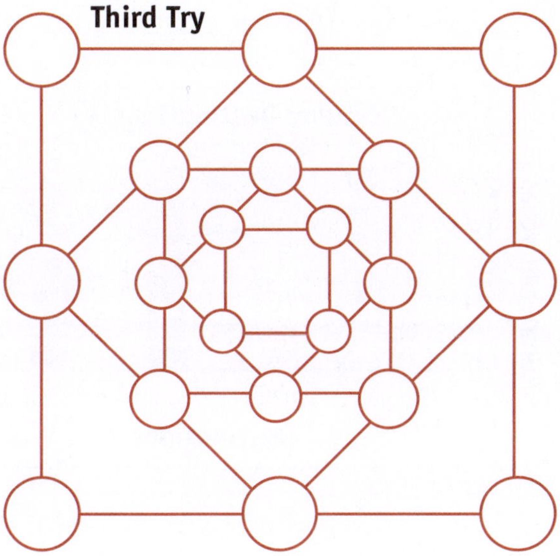

Fourth Try

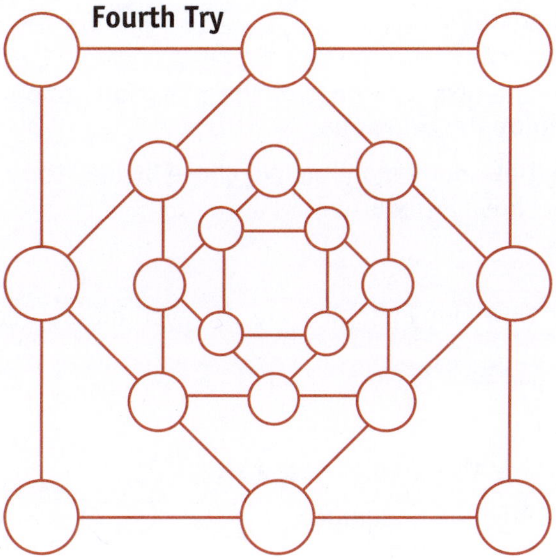

(3) Do you think there is a pattern? ______

If so, what do you think it is?

____

TARGETING MATHS HOMEWORK: YEAR 4 © PASCAL PRESS ISBN: 9781925726572

Review: Unit 17 and Unit 18

AC9M4N01, AC9M4N03, AC9M4N05, AC9M4N07, AC9M4A02, AC9M4M03, AC9M4SP01, AC9M4ST01

Write the value.

1. five  = ______
2. fourteen  = ______
3. seven  and fifteen  = ____________
4. twelve  = ______
5. five  = ______

Write these fractions in order from smallest to largest.

6. $1\frac{1}{8}$, $\frac{3}{8}$, $\frac{7}{4}$, $1\frac{1}{4}$, $\frac{3}{4}$

   ____

7. $1\frac{1}{2}$, $\frac{5}{4}$, $\frac{9}{8}$, $\frac{4}{4}$, $\frac{4}{2}$

   ____

Use the timelines to work out the answers. Answer in hours and minutes.

8. A movie starts at 7:10 pm, and ends at 9:05 pm. How long is the movie?

   Total time = ____________

9. The casserole went into the oven at 4:15 pm and was taken out of the oven at 6 o'clock. How long did the casserole cook for?

   Total time = ____________

Complete these multiplications.

10. 4 × 70 = ______
11. 5 × 50 = ______
12. 9 × 80 = ______
13. 9 × 7 = ______
14. 12 × 6 = ______
15. 16 × 4 = ______

Use the graph to answer the questions about holiday destinations.

16. How many more people chose Sydney than chose Melbourne?

    ____

17. How many people were surveyed altogether?

    ____

What is the area of these shapes?

18. Area = ____ squares

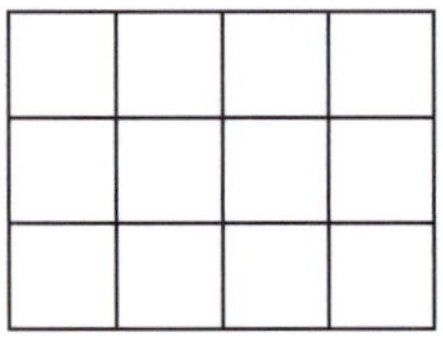

19. Area = ____ squares

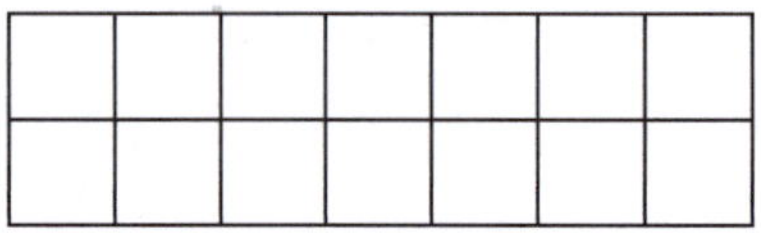

SCORE /19 0-9 10-17 18-19

TARGETING MATHS HOMEWORK: YEAR 4 © PASCAL PRESS ISBN: 9781925726572

Review: Unit 19 and Unit 20

AC9M4N01, AC9M4N05, AC9M4N07, AC9M4A01, AC9M4A02, AC9M4M03, AC9M4M04, AC9M4SP03, AC9M4P02

Complete these divisions.

(1) 720 ÷ 10 = ______ (3) 490 ÷ 10 = ______ (5) 540 ÷ 6 = ______

(2) 560 ÷ 7 = ______ (4) 450 ÷ 9 = ______ (6) 320 ÷ 4 = ______

Complete these additions with unknowns.

(7) 4 + ☐ = 23 (8) 17 + 236 = ☐ (9) ☐ + 28 = 57

(10) **Jo dropped a coin three times. How likely is it that her coin landed heads up twice?**

◯ Impossible ◯ Unlikely ◯ Likely ◯ Very likely ◯ Certain

Mark any lines of symmetry on these shapes.

(11)
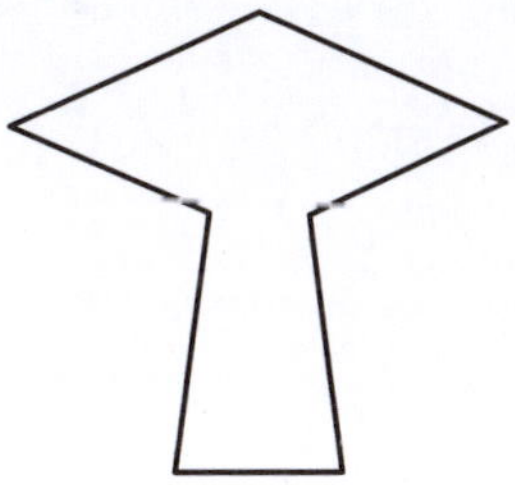

(12)

(13)
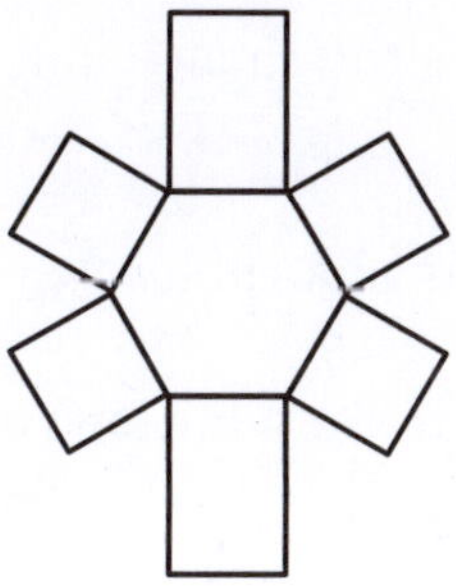

Round these numbers to the nearest whole number.

(14) 6·9 rounds to _____. (15) 4·3 rounds to _____. (16) 5·5 rounds to _____.

Round these numbers to the nearest tenth.

(17) 3·76 rounds to _____. (18) 2·83 rounds to _____. (19) 3·99 rounds to _____.

Complete these subtractions with unknowns.

(20) 36 − 19 = ☐ (21) ☐ − 29 = 25 (22) 94 − ☐ = 72

Name the type of angle made by the hour and minute hands.
Then write the time in words and digital form.

(23)

____________ angle

____

(24)

____________ angle

____

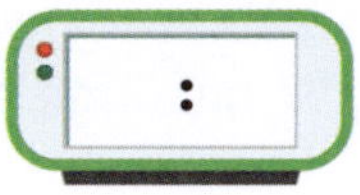

(25)

____________ angle

____

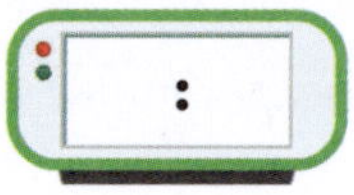

SCORE /25 0-12 13-23 24-25

Review: Unit 21 and Unit 22

AC9M4N03, AC9M4N05, AC9M4N08, AC9M4A01, AC9M4P01, AC9M4P02

Complete the equivalent fractions.

① $\frac{4}{8} = \frac{1}{\_\_}$ ② $\frac{4}{6} = \frac{\_\_}{3}$ ③ $\frac{4}{10} = \frac{2}{\_\_}$ ④ $\frac{3}{4} = \frac{6}{\_\_}$

Use the diagrams to solve these money problems.

⑤ $17 + ☐ = $45 ⑥ ☐ + $15 = $72 ⑦ $56 – ☐ = $19

Claire pulled two red counters and two blue counters out of the bag.

⑧ What is the chance that the next one will be black?

◯ Impossible ◯ Unlikely ◯ Likely ◯ Very likely ◯ Certain

Explain your choice. ______________________________

Complete these additions using the front-end method.

⑨ 28 + 69 = ______

⑩ 79 + 36 = ______

⑪ 126 + 364 = ______

⑫ 653 + 374 = ______

Round to the nearest dollar to estimate the total. Then check by finding the actual total.

	Round	Estimate	Actual
⑬ 5 sodas at $1·95 each.	______	______	______
⑭ 3 hot dogs at $2·85 each.	______	______	______

⑮ There were eight kittens in the litter.
What is the chance that there were four male kittens and four female kittens?

◯ Impossible ◯ Unlikely ◯ Likely ◯ Very likely ◯ Certain

⑯ How much fringe is needed for Claire's birthday cake which is an 8 cm square?

______ × ______ = ______

⑰ What is the perimeter of this book which is 297 mm high and 210 mm wide?

______ + ______ = ______

______ × ______ = ______

SCORE /17 0-8 9-15 16-17

TARGETING MATHS HOMEWORK: YEAR 4 © PASCAL PRESS ISBN: 9781925726572

Review: Unit 23 and Unit 24

AC9M4N03, AC9M4N09, AC9M4M01, AC9M4SP03

Follow the pattern rules to continue these sequences.

1. Multiply by 2, add 3.

   6, ____, ____, ____, ____

2. Subtract 2, multiply by 3.

   8, ____, ____, ____, ____

3. Divide by 2, add 6.

   68, ____, ____, ____, ____

4. Multiply by 4, subtract 3.

   6, ____, ____, ____, ____

Complete these decimal additions.

5. 2·4 + 3·4

   ____ ____·____

   ____·____

6. 3·7 + 4·4

   ____ ____·____

   ____·____

7. 4·6 + 5·9

   ____ ____·____

   ____·____

Find the area of these shapes.

8. 

   ____ × ____ = ____ squares

9. 

   ____ × ____ = ____ squares

Label these diagrams with the correct number sentence.

10. 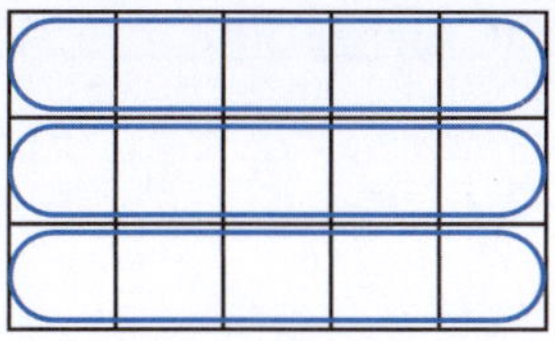

    ___ × ___ = ___

11. 

    ___ ÷ ___ = ___

12. 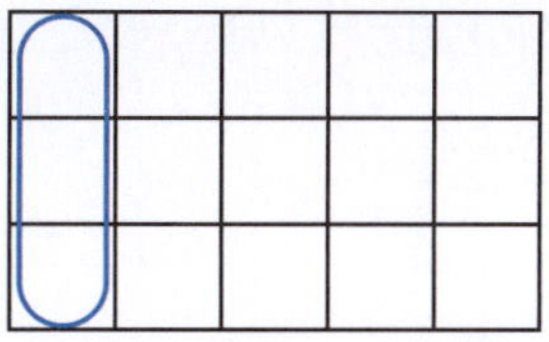

    $\frac{1}{\_\_}$ of ___ = ___

Write two division sentences for these multiplications.

13. 7 × 4

    ___ ÷ ___ = ___ and ___ ÷ ___ = ___

14. 9 × 6

    ___ ÷ ___ = ___ and ___ ÷ ___ = ___

Write the difference between these pairs of numbers.

15. 26 and 34

    The difference is ___.

16. 48 and 82

    The difference is ___.

Dan found that 4 apples have a mass of 1 kg.

17. What will be the mass of 1 apple? ________
18. What will be the mass of 8 apples? ________
19. How many apples will there be in a 1·5 kg bag of apples? _____
20. How many apples will there be in a 750 g bag of apples? _____

SCORE /20 0-9 10-18 19-20

TERM 3

TARGETING MATHS HOMEWORK: YEAR 4 © PASCAL PRESS ISBN: 9781925726572

Rounding and Estimating for Addition

There are two choices when rounding to make an estimate of an addition.
Here's an example. To make an estimate of **36 + 35**, you could:

- Round both numbers **down to 30**. The first estimate is **30 + 30 = 60**.
- Round both numbers **up to 40**. The second estimate is **40 + 40 = 80**.

The **estimated range** for the total is **between 60 and 80**.

Round each of the numbers in these additions to find an estimated range for the total.

(1) 28 + 52

____ + ____ = _____, ____ + ____ = _____

The range is between _____ and _____.

(2) 47 + 65

____ + ____ = _____, ____ + ____ = _____

The range is between _____ and _____.

(3) 76 + 54

____ + ____ = _____, ____ + ____ = _____

The range is between _____ and _____.

(4) 95 + 63

____ + ____ = _____, ____ + ____ = _____

The range is between _____ and _____.

(5) 75 + 62

____ + ____ = _____, ____ + ____ = _____

The range is between _____ and _____.

(6) 26 + 35 + 43

____ + ____ + ____ = _____

____ + ____ +____ = _____

The range is between _____ and _____.

(7) 92 + 35 + 48

____ + ____ + ____ = _____

____ + ____ +____ = _____

The range is between _____ and _____.

(8) 37 + 22 + 106

____ + ____ + ____ = _____

____ + ____ +____ = _____

The range is between _____ and _____.

(9) 45 + 62 + 79

____ + ____ + ____ = _____

____ + ____ +____ = _____

The range is between _____ and _____.

(10) 73 + 122 + 56

____ + ____ + ____ = _____

____ + ____ +____ = _____

The range is between _____ and _____.

SCORE /10 0-4 5-8 9-10

Quick Check

Give the estimated range for these additions.

(1) 43 + 57 _____ to _____

(2) 73 + 86 _____ to _____

(3) 54 + 75 _____ to _____

(4) 97 + 83 _____ to _____

(5) 85 + 87 _____ to _____

(6) 63 + 79 _____ to _____

(7) 27 + 32 _____ to _____

(8) 96 + 55 _____ to _____

(9) 176 + 48 _____ to _____

(10) 194 + 126 _____ to _____

SCORE /10 0-4 5-8 9-10

TARGETING MATHS HOMEWORK: YEAR 4 © PASCAL PRESS ISBN: 9781925726572

Number & Algebra

AC9M4N07

Rounding and Estimating Money Amounts

When I go shopping, I always make an estimate of my spending.
That way, I know I will have enough money to pay for my items. Here's an example.

I have three items in my trolley which cost **$1·75**, **$3·44** and **$7·20**. Altogether, that's a cost of **$12·39**.

Rounding to the nearest dollar value gives **$2 + $3 + $7 = $12**, which is the first estimate. This is a bit low because I rounded two of the amounts down, not up.

I now round one of those amounts up to make a second estimate: **$3 + $3 + $7 = $13**. If I have more than **$13**, I will be safe.

Round the amounts to find a safe estimate for the total.

1. $3·25 + $2·95 + $4·15

   ____ + ____ + ____ = ____

   A safe estimate is ____.

2. $4·75 + $3·45 + $6·70

   ____ + ____ + ____ = ____

   A safe estimate is ____.

3. $11·25 + $6·75 + $3·45

   ____ + ____ + ____ = ____

   A safe estimate is ____.

4. $15·70 + $3·40 + $12·25

   ____ + ____ + ____ = ____

   A safe estimate is ____.

5. $9·25 + $3·45 + $17·60

   ____ + ____ + ____ = ____

   A safe estimate is ____.

6. $27·35 + $6·95 + $5·50

   ____ + ____ + ____ = ____

   A safe estimate is ____.

SCORE /6 0-2 3-4 5-6

Statistics & Probability

AC9M4P01

Under the Cups

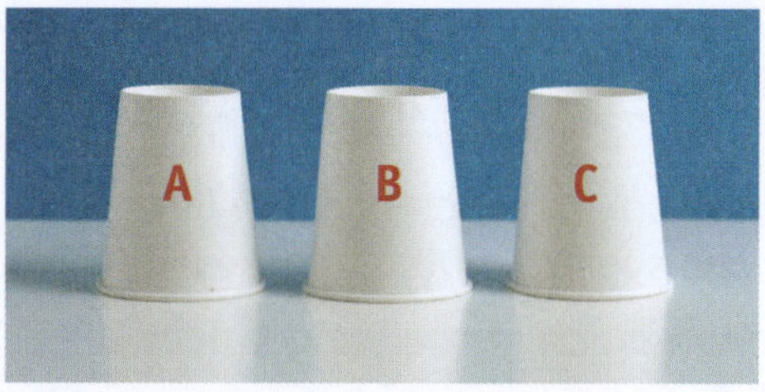

Claire has hidden 6 coins under these three cups.

1. Predict how many coins will be under cup B.

   ________ coins

2. Claire lifted Cup A and there were 2 coins under it.
   Predict how many coins will be under Cup B now that you know what is under cup A.

   ________ coins.

3. Explain your reasoning.

   ____

4. Claire tips Cup C just enough for you to see 1 coin under it.
   How many coins do you now think will be under Cup B?

   ________ coins

5. Explain your reasoning.

   ____

SCORE /5 0-2 3-4 5

TERM 4

AC9M4SP01

All About Cylinders

Many everyday objects are either cylinders or have a cylinder as part of their shape, like this bottle.

Cylinders are quite easy to draw.

First draw two ovals like this:

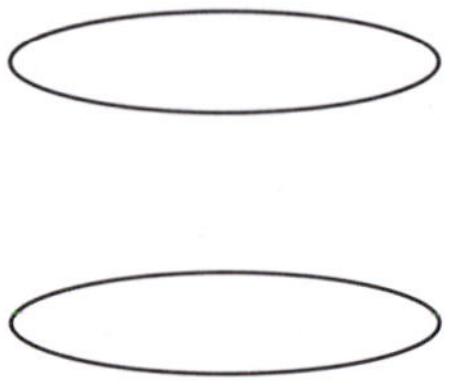

Then draw the sides like this:

You can erase the inside part of the bottom oval if you want to.

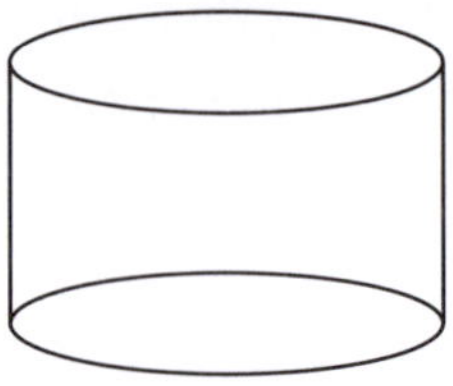

1. Draw two different cylinders in the boxes. (Score 2 points.)

2. List at least 8 objects in your house with a cylinder shape or which have a cylinder shape as part of them. (Score 8 points.)

SCORE /10 0-4 5-8 9-10

NAPLAN*-style Questions

1. The sports coach had $3478 to buy sports equipment.
   She spent $2000 on hockey equipment to set up a new hockey squad.
   How much of the budget does the coach have left?
   - ◯ $5478
   - ◯ $2478
   - ◯ $1478
   - ◯ $3478

2. Which two solid shapes were used to make this wooden kitchen roll holder?
   - ◯ cylinder
   - ◯ cone
   - ◯ rectangular prism
   - ◯ cube

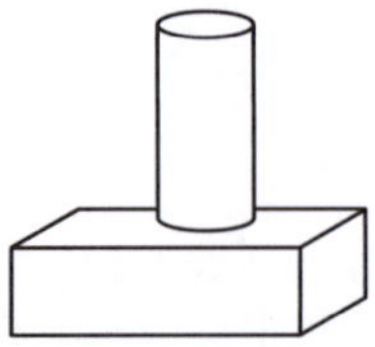

3. Claire rolled a dice six times.
   Which of these results is least likely?
   - ◯ 3 sixes
   - ◯ 6 sixes
   - ◯ 1 six

SCORE /3 0-1 2 3

TARGETING MATHS HOMEWORK: YEAR 4 © PASCAL PRESS ISBN: 9781925726572

Problem of the Week

Different Differences

1 5 8

Chen said, "I thought up three numbers – **1**, **5** and **8** – and then noticed that the differences between any two of them were all different."

She then explained what she meant:

"The difference between 1 and 5 is **4**, the difference between 1 and 8 is **7** and the difference between 5 and 8 is **3**.

All the differences are all different!"

Kai said, "If you want different differences, your largest number could be less than 8.

In fact, there are quite a few ways you can do it where the largest number is less than 8.

You should list the numbers you are using and their differences. That way you'll find more possibilities."

Find more different differences. Use three numbers.
The differences between any two of the numbers should be different and the largest number should be as small as possible.

Method 1	Method 2	Method 3

"What if you used **four numbers**?" asked Mrs Green, their teacher.

"We want the differences between any two of them to be different AND we want the largest number to be as small as possible."

Method 4	Method 5	Method 6

TARGETING MATHS HOMEWORK: YEAR 4 © PASCAL PRESS ISBN: 9781925726572

Estimating Money Amounts

When planning a budget, we usually overestimate to make sure that we can afford what we want.

To buy 9 boxes of apples at $19·40 a box, we could:

- use 10 boxes for the estimate because we know how to multiply by 10.
  10 × $19·40 = $194
- use 9 boxes at $20 because we can double 9 and multiply by 10.
  2 × 9 × 10 = $180

Both $194 and $180 are overestimates but $180 is the closer to the actual cost, which is **$174·60**.

Write two estimates for these amounts and then check to see which estimate is better. Tick the better estimate.

① 8 at $24·50

8 at $25 = ________

10 at $24·50 = ________

Actual cost: $________

② 9 at $15·80

___ at $________ = ________

___ at $________ = ________

Actual cost: $________

③ 19 at $3·60

___ at $________ = ________

___ at $________ = ________

Actual cost: $________

④ 11 at $22·50

___ at $________ = ________

___ at $________ = ________

Actual cost: $________

⑤ 12 at $14·35

___ at $________ = ________

___ at $________ = ________

Actual cost: $________

⑥ 28 at $16·25

___ at $________ = ________

___ at $________ = ________

Actual cost: $________

SCORE /5 0-2 3-4 5

Quick Check

Multiply by 10.

① $15·35 ________

② $27·15 ________

③ $46·05 ________

④ $17·36 ________

⑤ $26·55 ________

Round up and multiply by 20.

⑥ $13·75 ________

⑦ $15·55 ________

⑧ $32·80 ________

⑨ $43·60 ________

⑩ $37·85 ________

SCORE /10 0-4 5-8 9-10

TARGETING MATHS HOMEWORK: YEAR 4 © PASCAL PRESS ISBN: 9781925726572

Number & Algebra

AC9M4N09

Rules for Divisibility

÷2 If a number is divisible by 2 then it will be an even number. The digit in the ones place will be 0, 2, 4, 6 or 8.

÷3 If a number is divisible by 3 then its digits will **add to a multiple of 3**. For example, 7 × 3 = 21 and 2 + 1 = 3.

÷5 If a number is divisible by 5 then the digit in the ones place will be **either 0 or 5**.

÷9 If a number is divisible by 9 then its digits will **add to a multiple of 9**. For example, 7 × 9 = 72 and 7 + 2 = 9.

Use the rules for divisibility and tick which numbers are divisible by 2, 3, 5 or 9.

		Divisible by 2	Divisible by 3	Divisible by 5	Divisible by 9
1	6	✓			
2	30			✓	
3	27				
4	36				
5	55				
6	49				
7	56				
8	54				
9	42				
10	75				

SCORE /10 0-4 5-8 9-10

Statistics & Probability

AC9M4ST02

How Many People in Each State?

This table shows the population of five Australian States in 2021.

1 644 642	5 547 527	4 516 361	7 238 819	2 296 411
South Australia	Victoria	Queensland	New South Wales	Western Australia

Source: Australian Bureau of Statistics

1. Which state had the largest population? ____________
2. Which state had the smallest population? ____________
3. How many more people lived in Western Australia than in South Australia? ____________
4. How many fewer people lived in Victoria than in New South Wales? ____________
5. How many people altogether lived in those five states? ____________

SCORE /5 0-2 3-4 5

Measurement & Space

Timed Presentations

"Each group in our class is to make a 7-minute presentation. No more, no less." said Mrs Green. "You have time today for rehearsals and the presentations will be tomorrow."

At the rehearsals, each group timed themselves to work out how long their presentation was. Then they worked out whether they needed to add more material or make the presentation shorter, and if so, by how much.

Use the timelines to work out the lengths of the presentations. Then fill in the gaps and circle shorter or longer.

① Group 1 started at 11:55 and finished at 12:05.

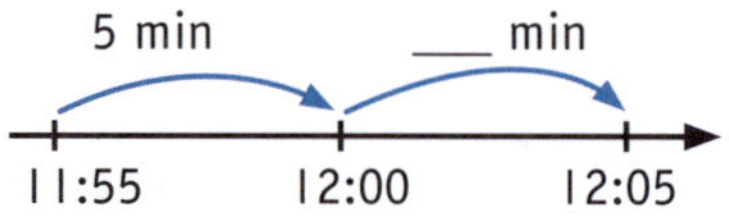

The presentation lasted ____ minutes and needs to be ____ minutes shorter / longer.

② Group 2 started at five to one and finished at 1:04.

The presentation lasted ____ minutes and needs to be ____ minutes shorter / longer.

③ Group 3 finished at 32 minutes past 2 after starting at 2:18.

The presentation lasted ____ minutes and needs to be ____ minutes shorter / longer.

④ Group 4 set a 30-second timer which went off 9 times.
They started at 2:42. At what time did they finish? ______________________

⑤ Work out the length of Group 4's presentation.

The presentation lasted ____ minutes and needs to be ____ minutes shorter / longer.

SCORE /5 (0-2) (3-4) (5)

NAPLAN*-style Questions

① Claire buys 3 sandwiches at $4·20 each.
How much change should she get from $20?

- ◯ $7·40
- ◯ $15·80
- ◯ $12·80
- ◯ $8·40

② Jo gets home at 3:50 and has 45 minutes of free time before she walks the dog.
At what time does Jo walk the dog? ▭

③ The train leaves Bridgeton at 11:15 am and arrives at Carlton at 1:45 pm.
How long does the journey take?

- ◯ 2 h 30 min
- ◯ 2 h
- ◯ 3 h 15 min
- ◯ 2 h 45 min

SCORE /3 (0-1) (2) (3)

TARGETING MATHS HOMEWORK: YEAR 4 © PASCAL PRESS ISBN: 9781925726572

Problem of the Week

Harshad Numbers

A **harshad number** is one that can be divided exactly by the sum of its digits.
For example, **12** is a Harshad number because **1 + 2 = 3** and **3** divides into **12** exactly.
45 is also a Harshad number.
Harshad is an Indian word meaning 'great joy'. We wish everyone great joy with this activity.

1. Circle the Harshad numbers you find on this grid.

0	1	2	3	4	5	6	7	8	9
10	11	12	13	14	15	16	17	18	19
20	21	22	23	24	25	26	27	28	29
30	31	32	33	34	35	36	37	38	39
40	41	42	43	44	45	46	47	48	49
50	51	52	53	54	55	56	57	58	59
60	61	62	63	64	65	66	67	68	69
70	71	72	73	74	75	76	77	78	79
80	81	82	83	84	85	86	87	88	89
90	91	92	93	94	95	96	97	98	99

2. Did you notice a pattern in where the Harshad numbers occur on the grid? ______

3. What is the pattern?

____

TARGETING MATHS HOMEWORK: YEAR 4 © PASCAL PRESS ISBN: 9781925726572

UNIT 27

Number and Algebra

AC9M4N02, AC9M4N06

Pattern Rules

Continue these pattern sequences and predict whether the 9th number will be odd or even. Here is an example.

+4, −2, +4, −2, +4, −2

1, 5, 3, 7, 5, 9, 7

All the numbers in this pattern are odd, so the 9th number will be **odd**.

① ____ ____ ____ ____ ____ ____

2, 5, 3, ◯, ◯, 7, ◯

The 9th number will be ________.

② ____ ____ ____ ____ ____ ____

3, ◯, 10, 20, 17, ◯, ◯

The 9th number will be ________.

③ ____ ____ ____ ____ ____ ____

◯, 30, 28, 25, ◯, ◯, ◯

The 9th number will be ________.

④ ____ ____ ____ ____ ____ ____

8, 16, ◯, 40, 44, ◯, ◯

The 9th number will be ________.

We would want to use a spreadsheet to find the 100th number in a pattern sequence.

Spreadsheets do not use the multiplication (×) and division (÷) signs, but use * for multiplication and / for division.

Also, spreadsheets use cell references like A2 and A3.

The pattern rule = **A3 + 2** means add 2 to the number in cell A2.

	A	B
1	Sequence	Pattern Rule
2	7	
3	9	=A2+2
4	45	=A3*5

Write the formulas for these pattern sequences.

⑤ Add 2, multiply by 5. = A2 + ___
= A3 * ___

⑥ Subtract 3, multiply by 6. = ________
= ________

⑦ Divide by 2, add 6. = ________
= ________

⑧ Multiply by 6, add 4. = ________
= ________

SCORE /8 0-3 4-6 7-8

TERM 4

Quick Check

Write the pattern rules for these sequences using the signs +, −, × and ÷.

① 3, 6, 8, 11, 13 ____, ____

② 2, 8, 6, 12, 10 ____, ____

③ 5, 15, 9, 27, 21 ____, ____

④ 29, 27, 37, 35, 45 ____, ____

⑤ 5, 6, 12, 13, 26 ____, ____

⑥ 1, 10, 110, 1100, 1200 ____, ____

⑦ 2, 6, 18, 22, 66 ____, ____

⑧ 7, 17, 14, 24, 21 ____, ____

⑨ 8, 2, 12, 6, 16 ____, ____

⑩ 9, 90, 89, 890, 889 ____, ____

SCORE /10 0-4 5-8 9-10

TARGETING MATHS HOMEWORK: YEAR 4 © PASCAL PRESS ISBN: 9781925726572

Number & Algebra

AC9M4N08

Word Problems

Write the number sentence and the solution for each of these word problems. You can use a calculator to help.

1. Dad said he would give Jo double the amount of money she had at the end of the week. Jo made $5 every day walking the neighbour's dog. How much did Dad give Jo?

   ______________ $__________

2. Claire had $36 at the beginning of the school week. She bought the $3·50 lunch at school every day and earned $10 for emptying the bins each school day. How much money did Claire have at the start of the weekend?

   ______________ $__________

3. Jake got $40 pocket money but owed his sister $12 and his mum $15. How much did Jake have left after paying back what he owed?

   ______________ $__________

4. Kai wants a new bike which is $496. He has saved $96 so far and will get $40 each week. How many more weeks before Kai has enough money to buy the bike?

   ______________ ____ weeks

5. Mrs Green buys a $6·50 cup of coffee on the way to school each morning. How much does that cost her for a 10-week term?

   ______________ $__________

SCORE /5 0-2 3-4 5

Statistics & Probability

AC9M4ST03

Our Pets

This pie graph shows the results of a survey of the pets that the 24 students in our class have at home.

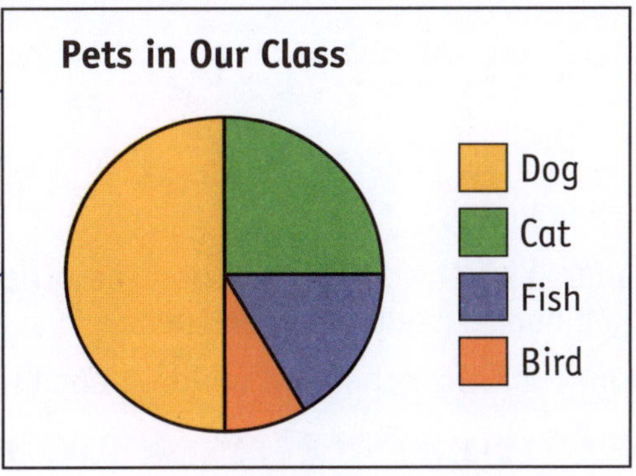

1. Which pet is the most popular? ______________
2. How many students have a dog at home? ______________
3. Do more students have a fish at home than a bird? ______________
4. How many students have a cat at home? ______________
5. What fraction of the class have a cat or a dog at home? ______________

SCORE /5 0-2 3-4 5

TERM 4

AC9M4SP02

Treasure Hunt

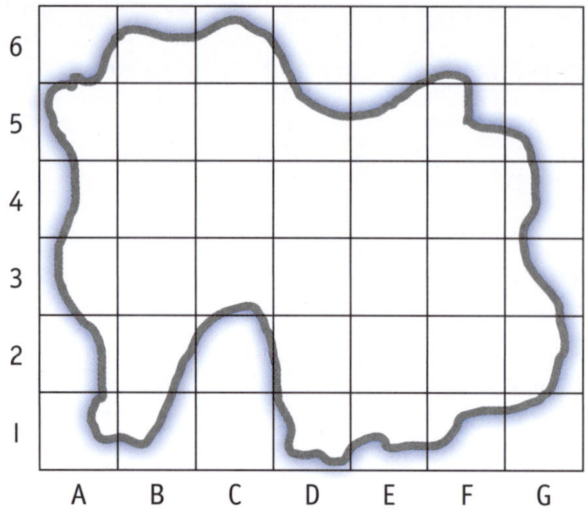

Draw the route to the treasure.

1. Start at D3 and walk 2 spaces to the east.
2. Turn 180° and walk forward 5 spaces.
3. Turn 90° north and walk forward 1 space.
4. Turn 360° and walk forward 1 space.
5. Turn 90° east and walk forward 2 spaces. You have reached the treasure.
6. What are the coordinates for the treasure? (___, ___)

SCORE /6 0-2 3-4 5-6

NAPLAN*-style Questions

1. What number is the arrow pointing to on this number line? [ ]

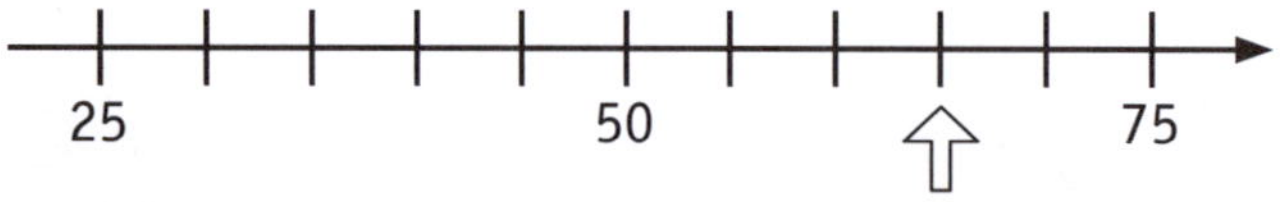

2. Four children shared $36.

   Two of the children received $12 each and the other two children shared the remainder equally between them.

   How much money did the other two children each receive?

   ◯ $9 ◯ $12 ◯ $6 ◯ $48

3. What are the map coordinates for the police station?

   (___, ___)

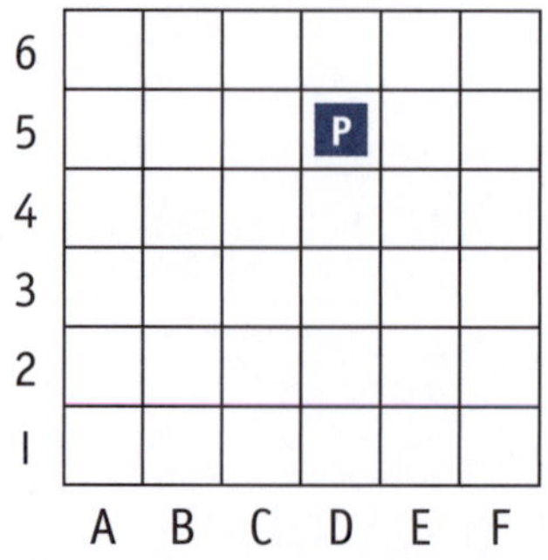

SCORE /3 0-1 2 3

TERM 4

TARGETING MATHS HOMEWORK: YEAR 4 © PASCAL PRESS ISBN: 9781925726572

Problem of the Week

AC9M4N09

Button Patterns

"Good morning, class. "As you know, I like to use my button collection to make up problems for you.

"Your problem for today," Mrs Green said, "is to find out where the buttons that are left in my button box belong in these button patterns.

All the buttons have a place, but you have to choose the last pattern to match the buttons that have not been used."

Fill in the missing numbers from the buttons in the box.
Then make a pattern with the leftover buttons.

+2, −1, +2, −1, +2, −1

1 → 3 → 2 → 4 → ___ → ___ → ___

___, ___, ___, ___, ___, ___

1 → ___ → ___ → ___ → 13 → 16 → 19

___, ___, ___, ___, ___, ___

___ → ___ → 8 → ___ → ___ → ___ → 128

___, ___, ___, ___, ___, ___

___ → ___ → ___ → ___ → ___ → ___ → ___

Here are the buttons that are left in the button box.

TARGETING MATHS HOMEWORK: YEAR 4 © PASCAL PRESS ISBN: 9781925726572

Finding Unknowns in Equivalent Number Sentences

To find the unknown in this number sentence: **16 + ☐ = 24 + 30**,
we first have to find 24 + 30, which is 54.
We then find the number that goes with 16 to make 54
so that both sides of the equals sign will balance.
That's 34. **16 + 34 = 24 + 30.**

Complete these equivalent number sentences, so that both sides of the equals sign balance.

1. 36 + 19 = ☐ + 26
2. 17 + 85 = 45 + ☐
3. ☐ + 16 = 48 + 4
4. 165 + ☐ = 179 + 8
5. 27 − 16 = 38 − ☐
6. 58 − ☐ = 67 − 14
7. 103 − 14 = ☐ − 99
8. ☐ − 44 = 103 − 4
9. 164 + ☐ = 201 − 31
10. 85 + ☐ = 65 + 75

Enter these key sequences on your calculator and record the results.
The first one has been done for you.

11. 23 [+] 2 [=] [×] 6 [=] 150
12. 3 [×] 6 [=] [÷] 2 [=] [+] 3 [=] ______
13. 2 [+] 6 [=] [÷] 4 [=] [×] 5 [=] ______
14. 49 [÷] 7 [=] [×] 3 [=] ______
15. 3 [×] 6 [=] [+] 2 [=] [÷] 5 [=] ______

SCORE /15 0-7 8-13 14-15

Quick Check

Complete these number sentences.

1. 38 + ☐ = 56
2. 59 + ☐ = 83
3. 76 − ☐ = 48
4. 83 − ☐ = 42
5. 76 + ☐ = 105
6. 26 + ☐ + 48 = 100
7. 38 + 54 − ☐ = 60
8. 37 − 14 + ☐ = 58
9. 29 + ☐ − 15 = 49
10. 27 − 2 + 16 = ☐

SCORE /10 0-4 5-8 9-10

TARGETING MATHS HOMEWORK: YEAR 4 © PASCAL PRESS ISBN: 9781925726572

Number & Algebra

UNIT 28

AC9M4N02, AC9M4N08

Make a Total Even, Make a Total Odd

Use the digits 1, 2, 3, 4, 5 and 6 to make three 2-digit numbers.
Then add the numbers together to make your total. But you can't use this example:

$$\begin{array}{r} 12 \\ 34 \\ +\ 56 \\ \hline 102 \end{array}$$

Make even totals here.

(1) + ______ ______

(2) + ______ ______

(3) + ______ ______

(4) + ______ ______

(5) + ______ ______

Make odd totals here.

(6) + ______ ______

(7) + ______ ______

(8) + ______ ______

(9) + ______ ______

(10) + ______ ______

SCORE /10 0-4 5-8 9-10

Statistics & Probability

AC9M4ST02, AC9M4ST03

Preparing for Camp

Mrs Green's class ran a survey to find out which drinks the class would prefer to take to camp. Here are the results.

Preferred Drinks

Orange juice	~~||||~~ ~~||||~~
Pineapple juice	||||
Apple juice	~~||||~~ ||
Lemon juice	~~||||~~

Base your answers on the survey results.

(1) Which juice wouldn't you take to camp?

______________ juice

Why? ______________________________

(2) Which juice would you take to camp?

______________ juice

(3) Four students were away on the day the survey was taken.
What is one way their votes could change your answer to Question 2?

____

Zoe is allergic to orange juice and Claire hates orange juice.
What other questions should the survey have included?

(4) ______________________________

(5) ______________________________

SCORE /5 0-2 3-4 5

TERM 4

AC9M4M01, AC9M4M04

Draw a House

Follow the instructions and label the points from A to G as you go. When completed, you will have drawn a nice modern house.

1. Start at A. Follow the line and mark a point B that is 10 cm from A.
2. Find the halfway point between A and B and label it C.
3. Draw an 8 cm line, starting at C and going up at right angles to AB. Label the end point D.
4. Mark the halfway point between C and D and label it E.
5. Draw a 3 cm line, starting at B and going up at right angles to AB. Label the end point F.
6. Join E to F with a straight line.
7. Draw a 7 cm line, starting at A and going up at right angles to AB. Label the end point G.
8. Join G to D with a straight line.

A

Add your own doors and windows!

SCORE /8 0-3 4-6 7-8

NAPLAN*-style Questions

1. Complete this number sentence.

   36 + ☐ = 59 − 13

2. Which of these lines is 3 cm long?

   ◯ ◯ ◯ ◯

3. Half of the 30 people surveyed had red cars. 6 had blue cars and the rest had silver cars.

   How many people had silver cars?

   ☐

SCORE /3 0-1 2 3

TERM 4

TARGETING MATHS HOMEWORK: YEAR 4 © PASCAL PRESS ISBN: 9781925726572

Problem of the Week

AC9M4N09

Making 20

Mrs Green showed this problem-solving challenge on the class screen:

Make every number from 1 to 20 using only these numbers and operations.

4 5 + −

Chen made 1 by making 16 and then subtracting 15:

4 + 4 + 4 + 4 − 5 − 5 − 5

Jo made 2 by making 12 and then subtracting 10:

4 + 4 + 4 − 5 − 5

Show how to make every number from 1 to 20.
But there is a challenge: make your expression to make 19 as short as possible.

1		11	
2		12	
3		13	
4		14	
5		15	
6		16	
7		17	
8		18	
9		19	
10		20	

AC9M4N05, AC9M4N06, AC9M4N08

The Safest Route

What is the safest route from 36·5 to 37 000?

Kai added 0·5 to make 36·5 into a whole number.

His method took 4 moves:

36·5 + 0·5 = 37

37 × 10 = 370

370 × 10 = 3 700

3700 × 10 = 37 000

Zoe first multiplied by 10 which removed the decimal point.

Her method took 3 moves:

36·5 × 10 = 365

365 × 100 = 36 500

36 500 + 500 = 37 000

1. **Show how to go from 36·5 to 730 in two moves.**

Show three safe routes to go from 6·4 to 64 500.

②	③	④

SCORE /4 0-1 2-3 4

Quick Check

Complete these safest route questions in no more than two moves each time.

1. From 3·5 to 400: 3·5 + 0·5 = 4
   4 × 100 = ________
2. From 3·6 to 3600: ________
3. From 5·7 to 600: ________
4. From 5·8 to 540: ________
5. From 6·8 to 7000: ________
6. From 4·9 to 5000: ________
7. From 6·05 to 61: ________
8. From 6·25 to 70: ________
9. From 7·75 to 75: ________
10. From 28·6 to 2860: ________

SCORE /10 0-4 5-8 9-10

TERM 4

TARGETING MATHS HOMEWORK: YEAR 4 © PASCAL PRESS ISBN: 9781925726572

Number & Algebra

AC9M4N08

Salad Plates

The kitchen team are making salad plates and they want all the plates to have exactly the same ingredients.

We have orders for **8 salad plates** and we do not want to waste anything.

We have:

- 32 radishes
- 6 tomatoes
- 48 cheese cubes.
- 4 avocados
- 24 spring onions

What should the chefs put on each plate?

1. ______________________
2. ______________________
3. ______________________
4. ______________________
5. ______________________

SCORE /5 0-2 3-4 5

Statistics & Probability

AC9M4P01

Pull Out a Red Counter

What is the chance of pulling a red counter out of each of these bags?

1. 
   - ◯ Impossible
   - ◯ Unlikely
   - ◯ Likely
   - ◯ Very likely
   - ◯ Certain

2. 
   - ◯ Impossible
   - ◯ Unlikely
   - ◯ Likely
   - ◯ Very likely
   - ◯ Certain

3. 
   - ◯ Impossible
   - ◯ Unlikely
   - ◯ Likely
   - ◯ Very likely
   - ◯ Certain

4. 
   - ◯ Impossible
   - ◯ Unlikely
   - ◯ Likely
   - ◯ Very likely
   - ◯ Certain

5. 
   - ◯ Impossible
   - ◯ Unlikely
   - ◯ Likely
   - ◯ Very likely
   - ◯ Certain

SCORE /5 0-2 3-4 5

AC9M4M03

Morning or Afternoon?

Mark am or pm for each of these statements.

1. Claire is going to the dentist at 1 o'clock.
   ◯ am ◯ pm
2. Jake is going to a sports training session at 10:30.
   ◯ am ◯ pm
3. Jo takes her dog for a walk every day before school.
   ◯ am ◯ pm
4. At midday, it is 12 o'clock.
   ◯ am ◯ pm
5. Claire and Jake are going star gazing at 9:30.
   ◯ am ◯ pm

Use the timelines to show how to find the hours between:

6. 11:00 pm to 11:00 am ____ hours
7. 3:00 am to 5:00 pm ____ hours
8. 2:00 pm to 6:00 pm ____ hours
9. 3:30 am to 1:30 pm ____ hours
10. 2:00 am to 8:00 pm ____ hours

SCORE /10 0-4 5-8 9-10

NAPLAN*-style Questions

1. Complete this number sentence: 37·5 + ☐ = 40·6
2. Jake went to the park at 11:20 am and stayed for three hours.
   Which time showns when Jake left the park?
   ◯ 2 o'clock pm ◯ 2:20 am ◯ 12:00 pm ◯ 2:20 pm
3. Which of these spinners is most likely to land on red?
   ◯ ◯ ◯ ◯

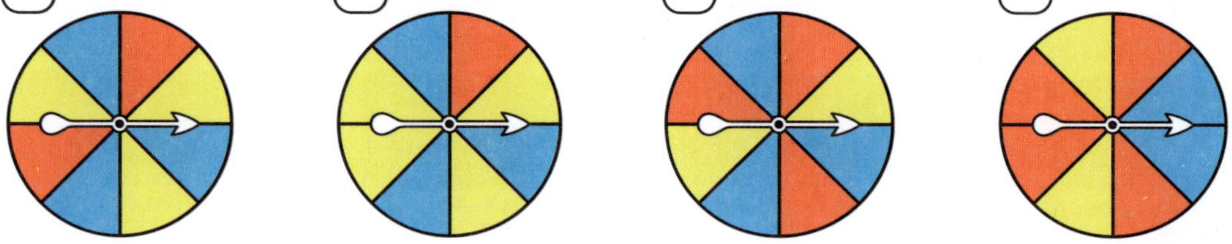

SCORE /3 0-1 2 3

TARGETING MATHS HOMEWORK: YEAR 4 © PASCAL PRESS ISBN: 9781925726572

Just by Chance

Oh dear! I am running late," said Mrs Green to her class, "and I am supposed to make up some bags of counters for the Lucky Draw section of the Maths Rally." She then showed the instructions for the Lucky Draw to her class.

Just by Chance – Lucky Draw

- **All bags have 12 counters.**

Bag 1 has to have the same chance of pulling out a blue, a red or a yellow counter.	**Bag 2** has to have an equal chance of pulling out a blue or red counter and very little chance of pulling out a yellow counter.
Bag 3 has to have an equal chance of pulling out a yellow or a red counter and an even higher chance of pulling out a blue counter.	**Bag 4** has to have no chance of pulling out a blue counter and a higher chance of pulling out a yellow counter than a red counter.

Colour the counters for each bag.

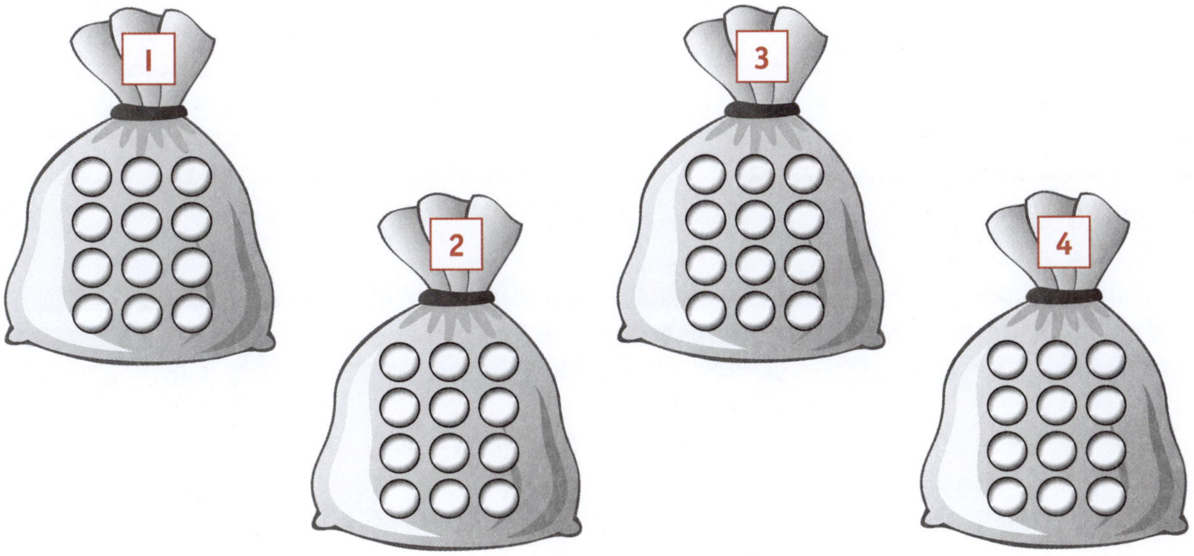

AC9M4N03

Fractions

Chef has these measuring cups and has set you some questions.

$\frac{1}{2}$ cup $\frac{1}{3}$ cup $\frac{1}{4}$ cup $\frac{1}{8}$ cup

I use my measuring cups every day in the kitchen, and I would like you to tell me the answers to these measuring problems.

1. If I needed 1 cup of flour using two cups, which cup should I use?
   The ________ cup
2. To measure 1 cup of flour using three cups all the same size, use the ________ cup.
3. To measure 1 cup of flour using three cups but only two cups the same size, use the ________ cup and ________ cup.
4. To measure 1 cup of flour using eight cups, use the ________ cup.
5. To measure 1 cup of flour using five cups, use the ____________________ cups.
6. To measure 1 cup of flour using four cups not all the same size, use the ____________________ cups.
7. If I needed 1 cup of flour and used seven cups not all the same size, I use ____________________ cups.
8. To measure $1\frac{1}{4}$ cups of flour and use five cups not all the same size, use the ____________________ cups.
9. To measure $1\frac{1}{2}$ cups of flour using six cups, use the ________ cup.
10. To measure $\frac{3}{4}$ of a cup of flour using four cups, use the ____________________ cups.

SCORE /10 0-4 5-8 9-10

Quick Check

1. $\frac{1}{2}$ cup + $\frac{1}{2}$ cup = ________
2. $\frac{1}{2}$ cup + $\frac{1}{4}$ cup = ________
3. $\frac{1}{8}$ cup + $\frac{1}{8}$ cup = ________
4. $\frac{1}{3}$ cup + $\frac{1}{3}$ cup = ________
5. $\frac{1}{4}$ cup + $\frac{1}{8}$ cup = ________
6. How many $\frac{1}{4}$ cups do you need to make $1\frac{1}{2}$ cups? ____
7. How many $\frac{1}{8}$ cups do you need to make a $\frac{1}{2}$ cup? ____
8. How many $\frac{1}{3}$ cups do you need to make 1 cup? ____
9. Is $\frac{1}{4}$ cup plus $\frac{1}{3}$ cup more or less than $\frac{1}{2}$ a cup? ________
10. How many $\frac{1}{8}$ cups do you need to make $\frac{3}{4}$ of a cup? ____

SCORE /10 0-4 5-8 9-10

Number & Algebra

AC9M4N08

Finding Halves and Quarters of Money Amounts

If we want to find **half of $3·20**, we need to think of $3·20 as **$2 + $1 + 20c** and then find halves.

Half of $2 = $1
Half of $1 = 50c
Half of 20c = 10c
} $1·60

To find **a quarter of $6**, we need to think of $6 as **$4 + $1 + $1** and then find quarters.

Quarter of $4 = $1
Quarter of $1 = 25c
Quarter of $1 = 25c
} $1·50

Or we might just know that $\frac{1}{4}$ of $2 is 50c and divide $6 into $4 + $2.

Find the fraction of these amounts.

1. Half of $7 = ____________________
2. Half of $3·60 = ____________________
3. Half of $27 = ____________________
4. Half of $105 = ____________________
5. Quarter of $8 = ____________________
6. Quarter of $9 = ____________________
7. Quarter of $6·80 = ____________________
8. Quarter of $17·60 = ____________________
9. Quarter of $120 = ____________________
10. Quarter of $24·40 = ____________________

SCORE /10 0-4 5-8 9-10

Statistics & Probability

AC9M4ST03

Cheaper by the Packet

This spreadsheet shows the cost of different-sized packets of three brands of orange juice.

	A	B	C	D
1	**Packet of**	**OJ**	**Big Orange**	**C-vit Orange**
2	1 carton	$3·45	$4·20	$3·85
3	2 carton	$6·45	$8·00	$7·40
4	3 carton	$9·90	$11·70	$10·80
5	4 carton	$13·00	$15·50	$14·40

1. What is the cost of a 3-carton packet of C-Vit Orange? __________
2. Which cell shows a cost of $11·70? __________
3. What does the amount in cell B4 give the cost of?

   ____
4. How much more expensive is it for 4 cartons of C-Vit Orange than for 4 cartons of OJ?

   ____
5. How much cheaper are 3 cartons of OJ than 3 cartons of Big Orange?

   ____

SCORE /5 0-2 3-4 5

TERM 4

AC9M4M02, AC9M4SP03

Symmetrical Shapes

Use the grid paper below to draw shapes with an area of 6 squares that have at least one line of symmetry. Here is one to start you off:

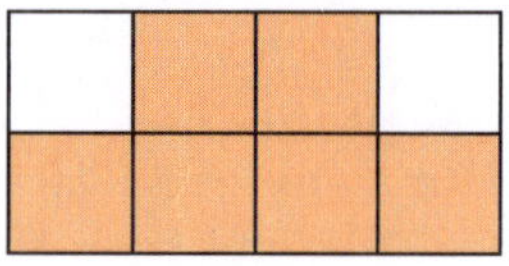

The squares must join along an edge:

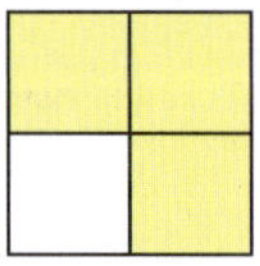

Not just at corners like this:

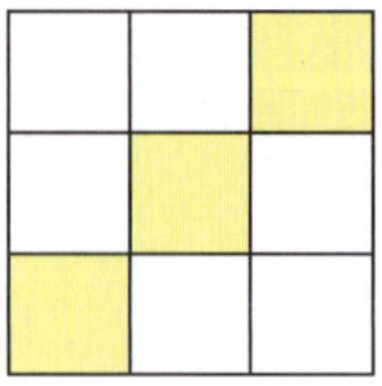

Mark any lines of symmetry that your shapes have.

Score 1 point for each symmetrical shape that you draw.

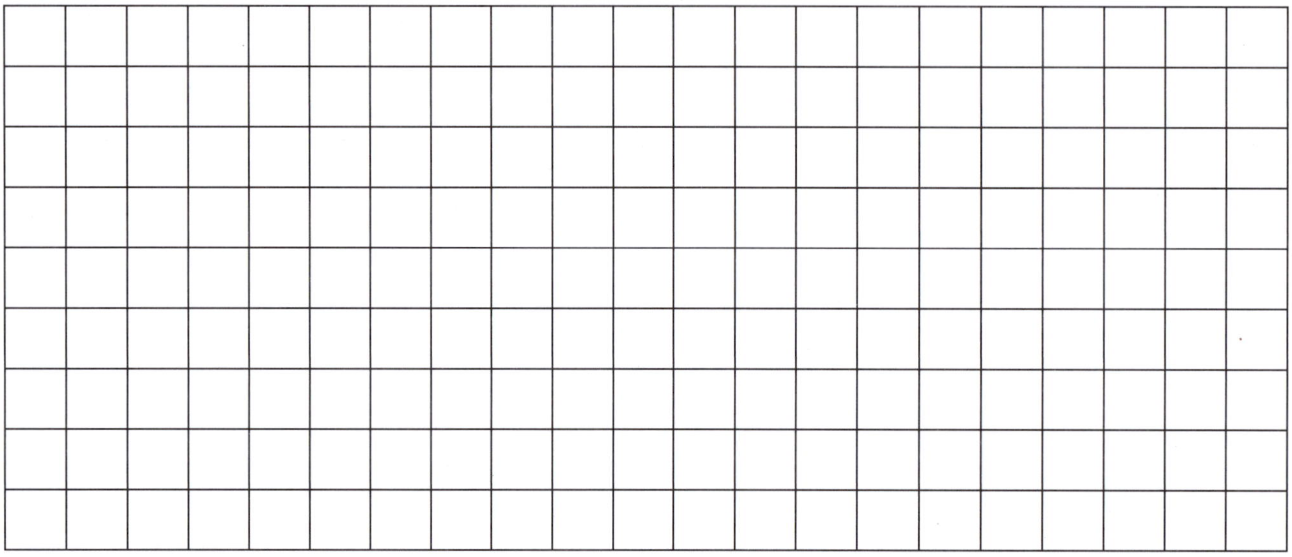

SCORE /10 0-4 5-8 9-10

NAPLAN*-style Questions

1. Which of these squares has the largest area that is unshaded?

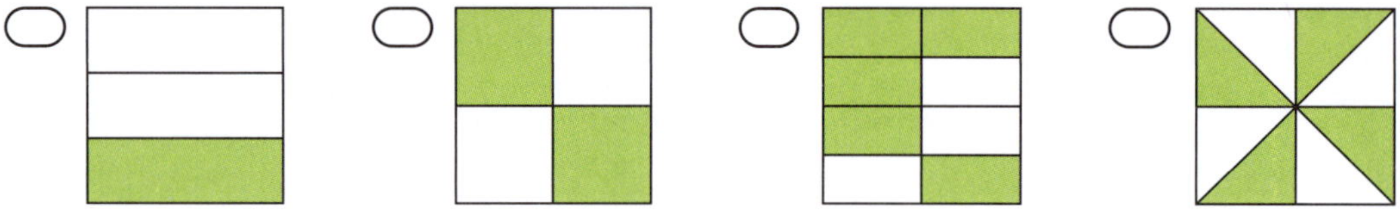

2. Claire, Jake, Jo and Chen each get $\frac{1}{4}$ of $50.

How much money is that?

$ ______

3. The spreadsheet shows the names of the players in three teams.

Which player's name is in cell C3?

______

	A	B	C
1	**Team 1**	**Team 2**	**Team 3**
2	Claire	Chen	Zoe
3	Jake	Kai	Paulo
4	Jo	Tim	Mia

SCORE /3 0-1 2 3

TERM 4

TARGETING MATHS HOMEWORK: YEAR 4 © PASCAL PRESS ISBN: 9781925726572

The Pizza Party

Last Saturday, Claire had a pizza party and invited six friends.
Her dad ordered four fancy pizzas and cut two into eighths and two into sixths.

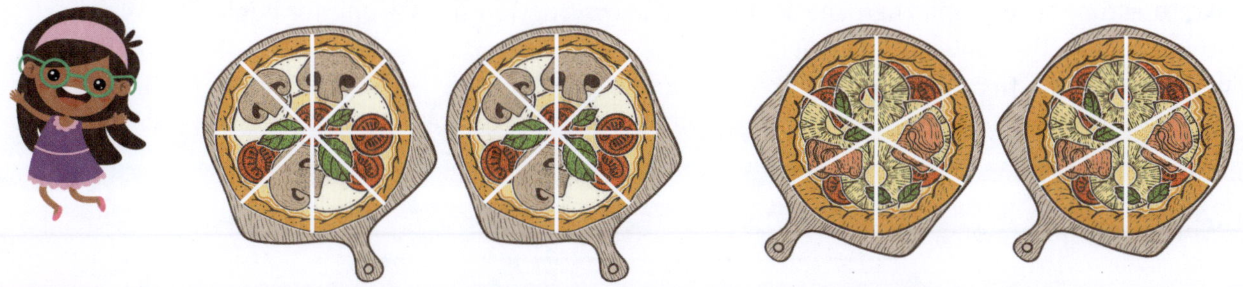

Zoe was late and there was only one small piece left for her.
Claire was upset because she had allowed half a pizza for each friend.
Someone had been greedy!

- Claire ate a quarter of one pizza and then a sixth of another.
- Jake ate half of one pizza and then a sixth of another.
- Jo ate an eighth of one pizza, a quarter of another and half of another.
- Chen ate an eighth of one pizza, a sixth of another and a quarter of another.
- Kai ate half of one pizza and then a sixth of another.
- Dan ate a third of one pizza and then a quarter of another.

Use these pizza bases to show who ate how much before Zoe arrived at the pizza party.

(1) What fraction of a pizza was left for Zoe? ______

(2) Who ate the most pizza? ______________

AC9M4N04

Comparing Fractions

You can use fraction lines to compare fractions when you need to know which of two fractions is larger.

I like to estimate first and then check by making two fraction lines – one for each of the fractions that you want to compare.

I have started the first one for you.

Compare the fractions.

1. Which is larger, $\frac{2}{3}$ or $\frac{3}{4}$?

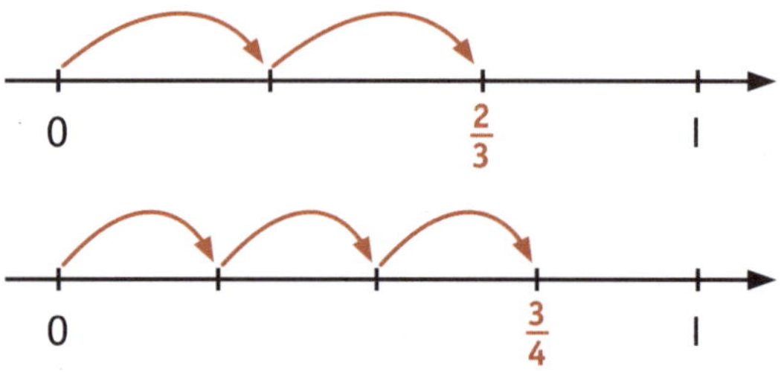

Estimate: ___ is larger.

Check: ___ is larger/smaller.

2. Which is larger, $\frac{3}{5}$ or $\frac{6}{8}$?

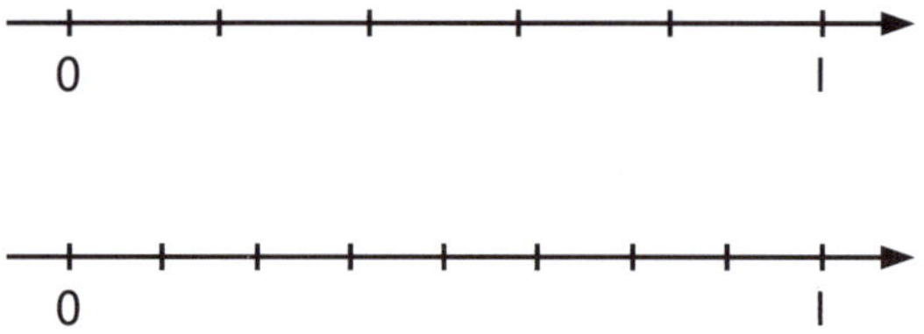

Estimate: ___ is larger.

Check: ___ is larger/smaller.

3. Which is larger, $\frac{6}{8}$ or $\frac{3}{4}$?

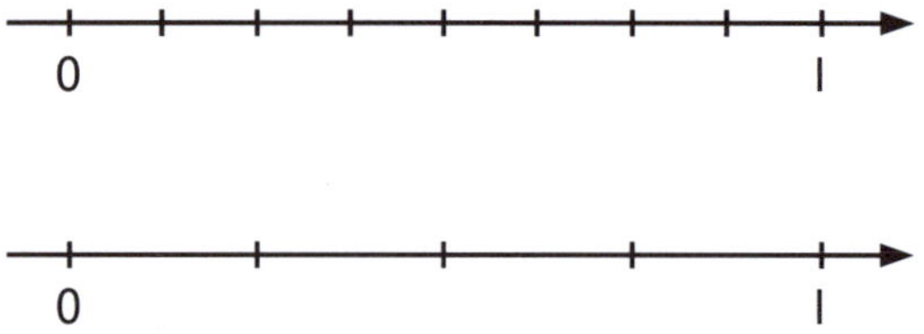

Estimate: ___ is larger.

Check: ___ is larger/smaller.

4. Which is larger, $\frac{7}{8}$ or $\frac{7}{10}$?

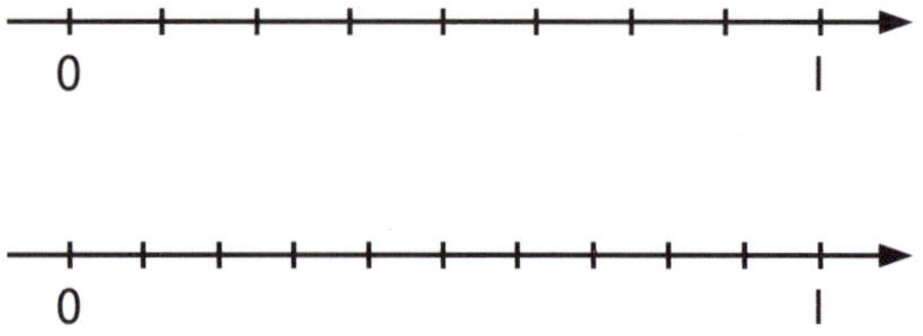

Estimate: ___ is larger.

Check: ___ is larger/smaller.

SCORE /4 | 0-1 | 2-3 | 4

TERM 4

Quick Check

Which is larger?

1. $\frac{1}{5}$ or $\frac{1}{6}$ ____
2. $\frac{1}{8}$ or $\frac{1}{3}$ ____
3. $\frac{2}{8}$ or $\frac{1}{4}$ ____
4. $\frac{3}{8}$ or $\frac{1}{2}$ ____
5. $\frac{1}{2}$ or $\frac{3}{6}$ ____

Which is smaller?

6. $\frac{1}{4}$ or $\frac{3}{8}$ ____
7. $\frac{3}{4}$ or $\frac{7}{8}$ ____
8. $\frac{5}{6}$ or $\frac{1}{2}$ ____
9. $\frac{1}{2}$ or $\frac{2}{3}$ ____
10. $\frac{7}{8}$ or $\frac{7}{5}$ ____

SCORE /10 | 0-4 | 5-8 | 9-10

TARGETING MATHS HOMEWORK: YEAR 4 © PASCAL PRESS ISBN: 9781925726572

Number & Algebra

AC9M4N08

Fractions and Division

If you divide this group of counters in half, you can write two number sentences.

$\frac{1}{2}$ **of 6 = 3** and **6 ÷ 2 = 3**

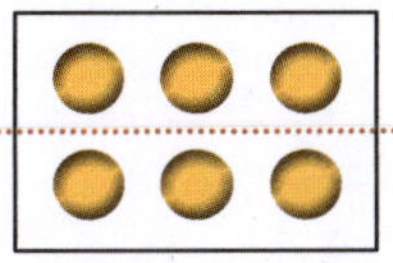

Complete these fraction diagrams and write two number sentences for each one.

1. Find $\frac{1}{4}$ of these counters.

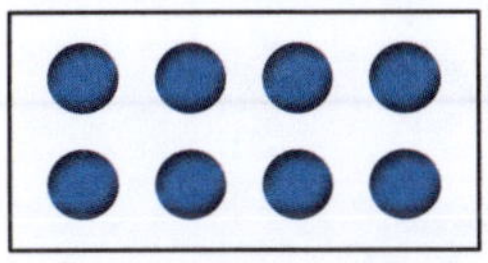

— of ☐ = ☐ and ☐ ÷ ☐ = ☐

2. Find $\frac{1}{5}$ of these counters.

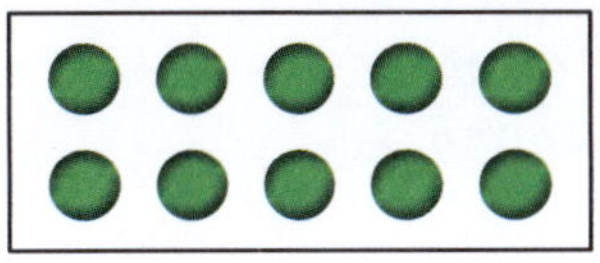

— of ☐ = ☐ and ☐ ÷ ☐ = ☐

3. Find $\frac{1}{6}$ of these counters.

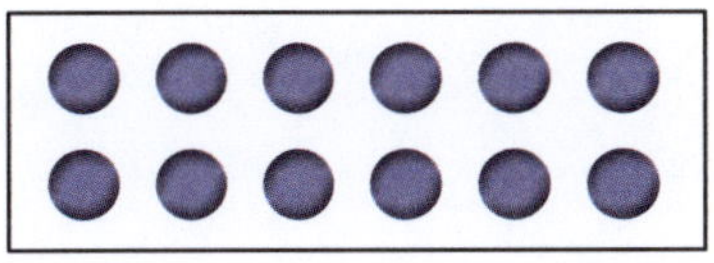

— of ☐ = ☐ and ☐ ÷ ☐ = ☐

4. Find $\frac{1}{3}$ of these counters.

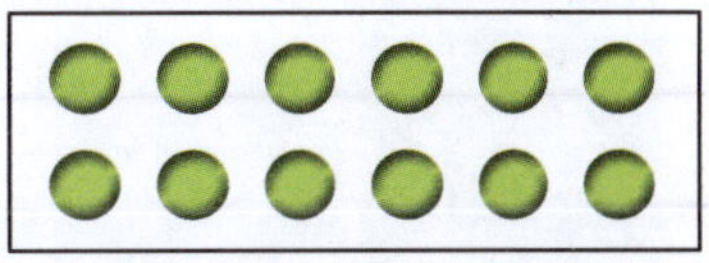

— of ☐ = ☐ and ☐ ÷ ☐ = ☐

5. Find $\frac{1}{8}$ of these counters.

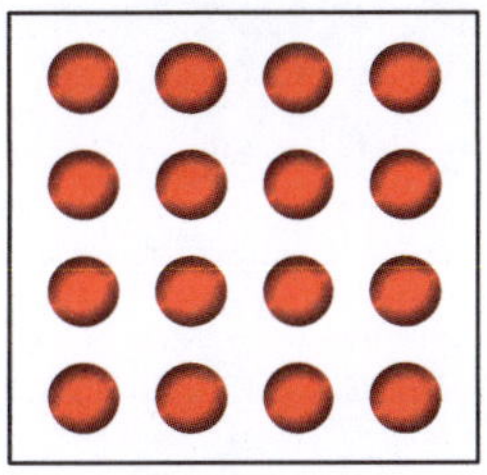

— of ☐ = ☐ and ☐ ÷ ☐ = ☐

SCORE /5 0-2 3-4 5

Statistics & Probability

AC9M4ST01

The Spellathon Results

After the Spellathon, the results were graphed in two ways.

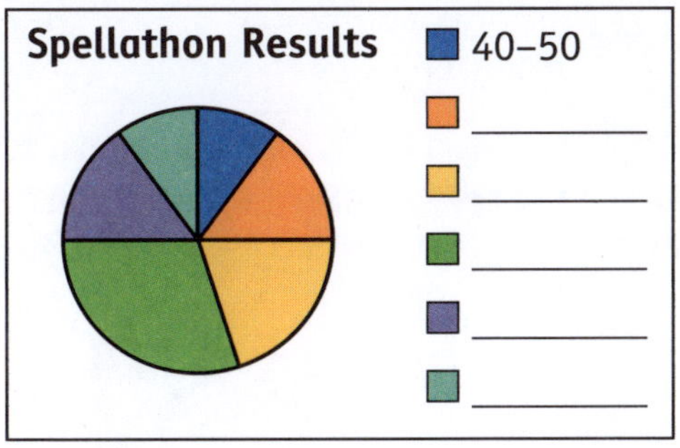

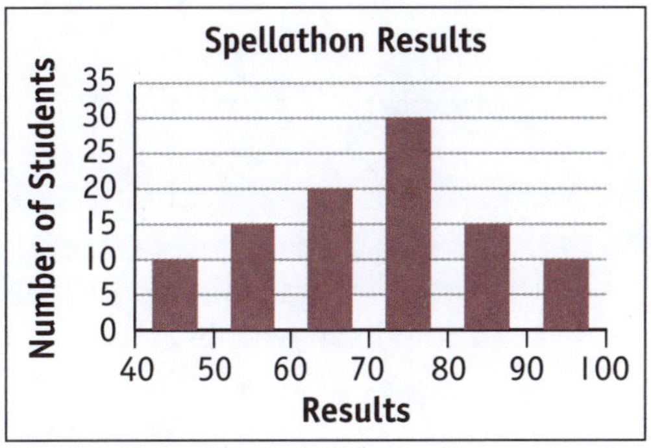

1. Complete the key for the pie graph to match the data in the bar graph.
2. How many students were in the spellathon altogether? _______ students
3. How many students scored more than 80? ______ students
4. How many students scored between 60 and 90? ______ students

SCORE /4 0-1 2-3 4

Measurement & Space

AC9M4M02

The Garden Design

Zoe's family is replacing part of the lawn in the back yard with the garden in this design.

Scale: ← 1 m →

1. How much fencing is needed to enclose this garden? ______ m
2. How many square metres of paving are needed for the garden path? ______ m^2
3. How many square metres is the chicken pen? ______ m^2
4. Special rabbit-proof fencing is needed for the vegetable patch. How much fence will it take to go all around the patch? ______ m
5. How many square metres is for fruit trees? ______ m^2
6. How much space does the garden shed take up? ______ m^2

SCORE /6 (0-2) (3-4) (5-6)

NAPLAN*-style Questions

1. Which rectangle has the same fraction shaded as the circle?

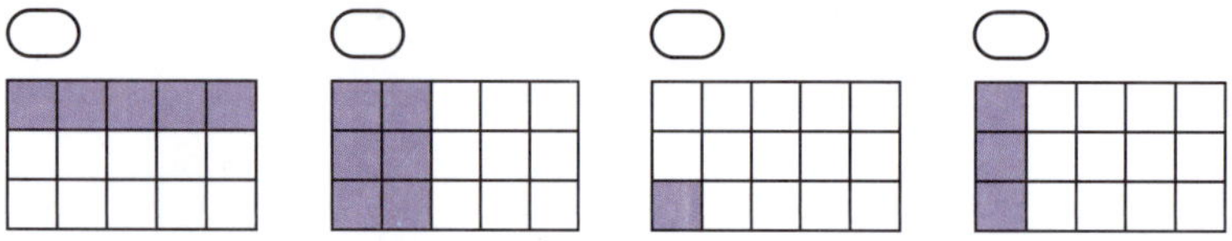

2. The sandwich bar sold more chicken sandwiches than salad sandwiches.

   It also sold more cheese than ham sandwiches and less chicken than ham sandwiches.

   Which column of the sales chart is for salad sandwiches?

   ◯ A ◯ B ◯ C ◯ D

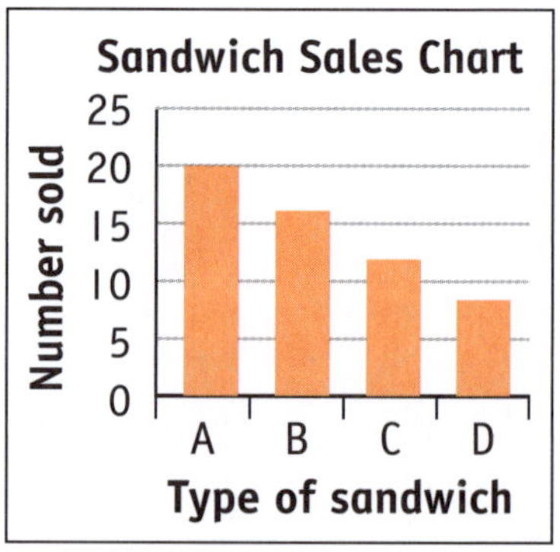

3. What is $\frac{1}{3}$ of 12? ☐

SCORE /3 (0-1) (2) (3)

TERM 4

TARGETING MATHS HOMEWORK: YEAR 4 © PASCAL PRESS ISBN: 9781925726572

Problem of the Week

AC9M4N08

Counters by Colours

Claire, Jake and Chen have been colouring counters and making up clues that let you work out how many counters there are of each colour.

I had eighteen counters and coloured $\frac{1}{6}$ of them red.
Then for every red counter I made two yellow ones and three green ones.

Colour these counters to find out how many I had of each colour.

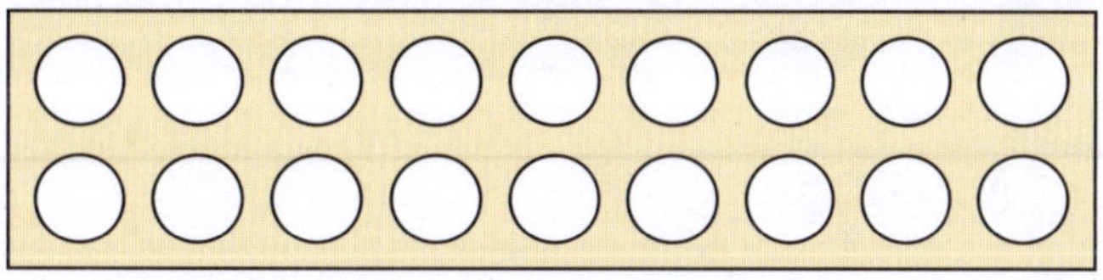

① Claire had _____ red, _____ yellow and _____ green counters.

I had twenty counters and $\frac{1}{4}$ of them are green.
For every green counter there are two red counters, and the rest are yellow.

Colour these counters to find out how many I had of each colour.

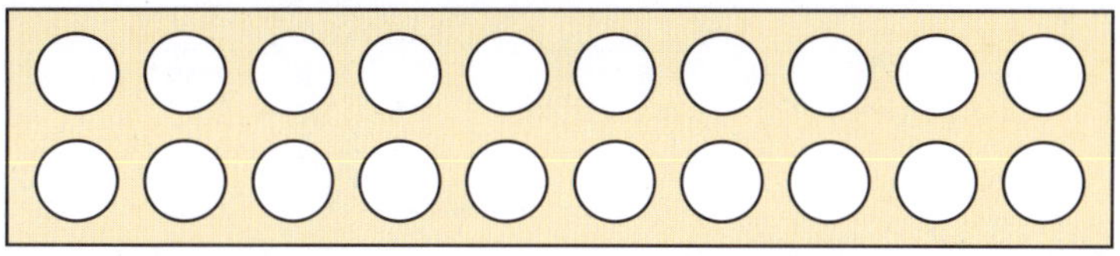

② Jake had _____ red, _____ yellow and _____ green counters.

I had thirty counters and $\frac{1}{5}$ of them are green.
For every yellow counter there are two green counters and three black counters.
For every green counter there are two red counters.

Colour these counters to find out how many I had of each colour.

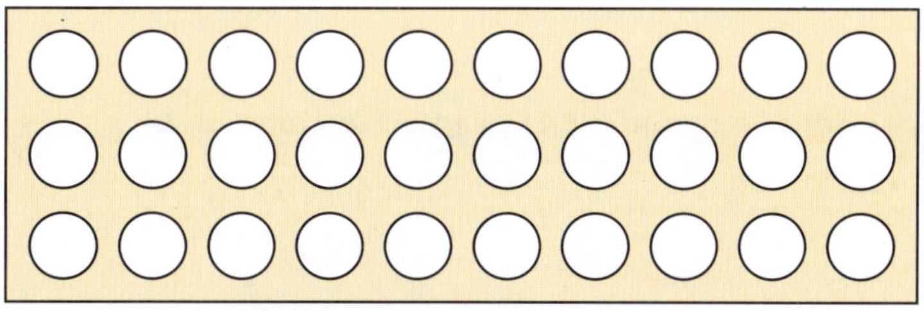

③ Chen had _____ red, _____ yellow, ____ green and ____ black counters.

④ Your turn now!
Use these counters to make your own mixed counter clues.

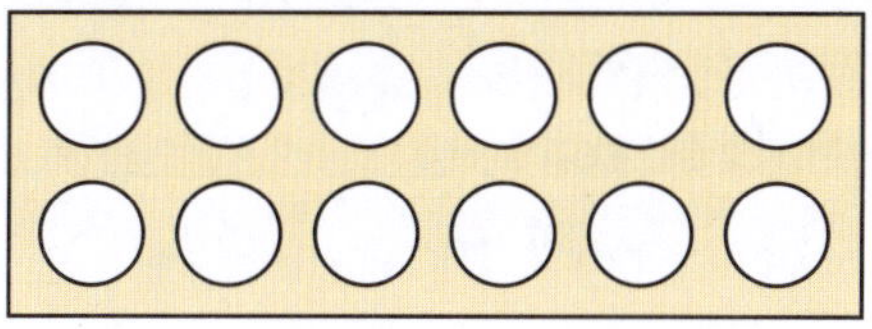

Write your clues here:

____

____

____

⑤ How many counters of each colour do you have?

____

Money Amounts

If I want to work out what **forty 10c coins** are worth, I use what I know about 10c coins.

I know that 10 × 10c makes $1.
So I divide the 40 by 10, which gives **40 ÷ 10 = 4**, which makes **$4**.

To find what **thirty 20c coins** are worth, I would use 5 × 20c = $1.
So I divide 30 by 5 which gives **30 ÷ 5 = 6**, which makes **$6**.

Try these examples.

1. For sixty 5c coins, think 20 × 5c = $1.

   _____ ÷ _____ = _____, which makes $_____.

2. For one hundred and twenty 10c coins, think _____ × 10c = $1.

   _____ ÷ _____ = _____, which makes $_____.

3. For twelve 50c coins, think _____ × 50c = $1.

   _____ ÷ _____ = _____, which makes $_____.

4. For eighty 20c coins, think _____ × 20c = $1.

   _____ ÷ _____ = _____, which makes $_____.

5. For eighty 5c coins, think _____ × 5c = $1

   _____ ÷ _____ = _____, which makes $_____.

6. For one hundred and forty 5c coins, think _____ × 5c = $1

   _____ ÷ _____ = _____, which makes $_____.

7. For fifty 50c coins, think _____ × 50c = $1

   _____ ÷ _____ = _____, which makes $_____.

8. For two hundred and twenty 10c coins, think _____ × 10c = $1

   _____ ÷ _____ = _____, which makes $_____.

SCORE /8 0-3 4-6 7-8

TERM 4

Quick Check

What is the value?

1. Forty 5c coins = __________
2. Twenty-eight 10c coins = __________
3. Fifty-five 20c coins = __________
4. Forty 50c coins = __________
5. Sixty $1 coins = __________
6. Thirty 50c coins = __________
7. Twenty-one 5c coins = __________
8. Eleven 10c coins = __________
9. Twenty-one 20c coins = __________
10. Eighty-five 20c coins = __________

SCORE /10 0-4 5-8 9-10

TARGETING MATHS HOMEWORK: YEAR 4 © PASCAL PRESS ISBN: 9781925726572

Number & Algebra

UNIT 32

AC9M4N09

Money Pattern Rules

Follow the pattern rules ten times to find out which one ends on the largest total.

① Start with 5c and double the amount each time.

5c, 10c, ______, ______, ______, ______, ______, ______, ______, ______

② Start with $6·50 and add on 20c each time.

$6·50, ______, ______, ______, ______, ______, ______, ______, ______, ______

③ Start with $2·05 and add on 50c each time.

______, ______, ______, ______, ______, ______, ______, ______, ______, ______

④ Start with 15c, then add on 10c. Double the amount being added on each time.

______, ______, ______, ______, ______, ______, ______, ______, ______, ______

⑤ Start with 55c and add on 5c, then 10c, then 20, then 50c, then 5c and keep that pattern going.

______, ______, ______, ______, ______, ______, ______, ______, ______, ______

⑥ The pattern that ends on the largest total is pattern ____.

SCORE /6 0-2 3-4 5-6

Statistics & Probability

AC9M4P02

Pick Out 40c

For this experiment, make 8 small scraps of paper and write these money amounts on them:

5c	5c	10c	10c	20c	20c	50c	50c

Put your pieces of paper into a bag and give it a shake.

When you take out 2 pieces of paper, record the amount on each one and put them back in the bag. Give the bag a shake.

① How many times do you think you will have to do the experiment before you take out 2 pieces worth more than 30c?

_____ times

② **Experiment Time**

Take out 2 pieces of paper and record the total value 10 times. Record your values and tick those that are above 30c.

Do not look as you take the pieces of paper from your bag, and make sure you give the bag a shake after you put the pieces back.

What do you now think is the chance of taking out more than 30c?

	Value	Above 30c
1		
2		
3		
4		
5		
6		
7		
8		
9		
10		

◯ Impossible ◯ Unlikely ◯ Likely ◯ Very likely ◯ Certain

③ If you repeat that experiment 10 more times, which value do you think the pieces of paper will make most often? ____ cents

Why is that? ______________________________

SCORE /3 0-1 2 3

AC9M4SP01

Shape Creatures

These shape creatures have been made using a combination of these shapes, which were drawn on a 1 cm^2 grid.

Isosceles triangle | Square | Rectangle | Right-angle triangle

1. Which shapes and how many of them were used to make the shape-creatures?

   (Score one point for each correct entry in the table.)

	Shape A	Shape B	Shape C
Isosceles triangle			
Square			
Rectangle			
Right-angle triangle			

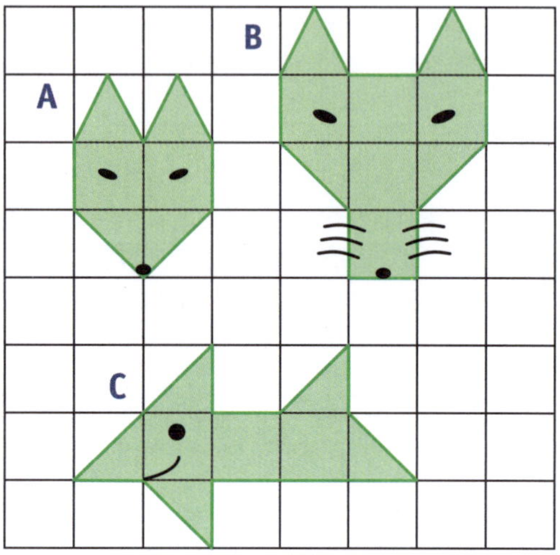

2. What is the area of Shape A?

   ______ squares

3. What is the area of Shape B?

   ______ squares

4. What is the area of Shape C?

   ______ squares

SCORE /15 0-7 8-13 14-15

NAPLAN*-style Questions

1. Claire spent $12·20 on party balloons. Each balloon cost 20c.
   How many balloons did Claire buy?

   [ ]

2. Without looking, Jake took a numbered counter from this bag.
   Which number is he most likely to have taken?

   ◯ 1 ◯ 2 ◯ 3 ◯ 4

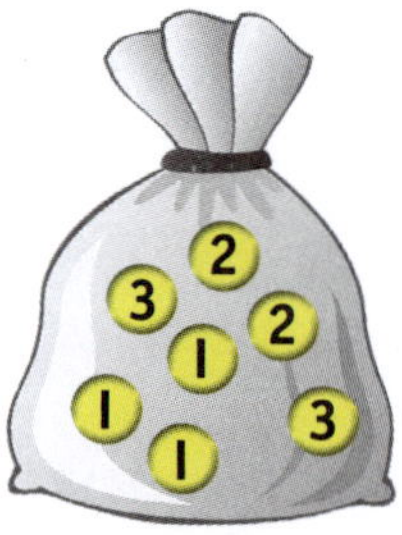

3. Which of these shapes is not used in the house picture?

   ◯ square ◯ rectangle
   ◯ semicircle ◯ triangle

SCORE /3 0-1 2 3

TERM 4

TARGETING MATHS HOMEWORK: YEAR 4 © PASCAL PRESS ISBN: 9781925726572

Problem of the Week

Coins in the Purse

Claire:

Mum lets me have any 5-cent and 20-cent coins that are left in her purse at the end of the week. I put them in a special jar.

By the time I had collected $4, I had 44 coins in my jar.

Jo:

My mum lets me have the 10c and 50c coins that are left in her purse. I too have a special jar where I keep the coins.

By the time I had collected $8, I also had 44 coins.

Can you work out what coins Claire and Jo collected?
Show your working here.

Claire's Coins	**Jo's Coins**

Two weeks later, Claire has 60 coins with a value of $6.
Jo also has 60 coins and the value is $12.
What coins do they both have now?

Claire's Coins	**Jo's Coins**

Review: Unit 25 and Unit 26

AC9M4N03, AC9M4N07, AC9M4M03, AC9M4SP01

Round each of the numbers in these additions to find an estimated range for the total.

(1) 36 + 23

____ + ____ = _____, ____ + ____ = _____

The range is between _____ and _____.

(2) 43 + 58

____ + ____ = _____, ____ + ____ = _____

The range is between _____ and _____.

(3) 74 + 88

____ + ____ = _____, ____ + ____ = _____

The range is between _____ and _____.

(4) 197 + 34

____ + ____ = _____, ____ + ____ = _____

The range is between _____ and _____.

(5) Draw a cylinder in the box.

Round the amounts to make a safe estimate for the total.

(6) \$3·75 + \$2·20 + \$3·55

____ + ____ + ____ = ____

A safe estimate is ____.

(7) \$16·45 + \$3·75 + \$12·20

____ + ____ + ____ = ____

A safe estimate is ____.

Make safe estimates of these amounts

(8) 8 apples at \$1·65 each

8 at \$_______ = _______

10 at \$_______ = _______

(9) 9 pencils at \$2·20 each

___ at \$_______ = _______

___ at \$_______ = _______

(10) Which of these numbers are divisible by 9?

◯ 12 ◯ 19 ◯ 27 ◯ 35 ◯ 45

(11) Which of these numbers are divisible by 3 and by 5?

◯ 9 ◯ 30 ◯ 15 ◯ 27 ◯ 55

(12) Which of these numbers are divisible by 3, 5 and 9?

◯ 35 ◯ 45 ◯ 55 ◯ 60 ◯ 90

(13) The train left the station at 9:45 am.
The journey lasted 3 hours and 30 minutes.
At what time did the train arrive at its destination? ___:____ am/pm

(14) The train left the station at 6:35 am and arrived at its destination at 12:05 pm.
How long was the journey? ____ h ___ min

SCORE /14 0-6 7-12 13-14

TARGETING MATHS HOMEWORK: YEAR 4 © PASCAL PRESS ISBN: 9781925726572

Review: Unit 27 and Unit 28

AC9M4N07, AC9M4N08, AC9M4M01, AC9M4ST03

Complete these pattern sequences.

① +4 −3 ____ ____ ____ ____

(6) ( ) ( ) (11) ( ) ( ) ( )

② ____ ____ ____ ____ ____ ____

(3) (6) ( ) ( ) ( ) (30) (31)

Complete these sequences and write their pattern rules.

③ (2) (4) ( ) (5) (4)

Rule: ____ and ____

④ (5) ( ) (9) (18) ( ) (34)

Rule: ____ and ____

Write the number sentence and the answer for each of these word problems.

⑤ Jake spent $4 every day this week.
He had $40 at the beginning of the week.
How much does he have now?

Number sentence: ____________________

Jake has $ ______ left.

⑥ Every day in June, Claire earned $5·50 for walking the dog.
How much did Claire earn in June?

Number sentence: ____________________

Claire earned $ ______ in June.

The class were allowed to choose the colour for their new desks. The graph shows their choices.

New Desk Colour Choices

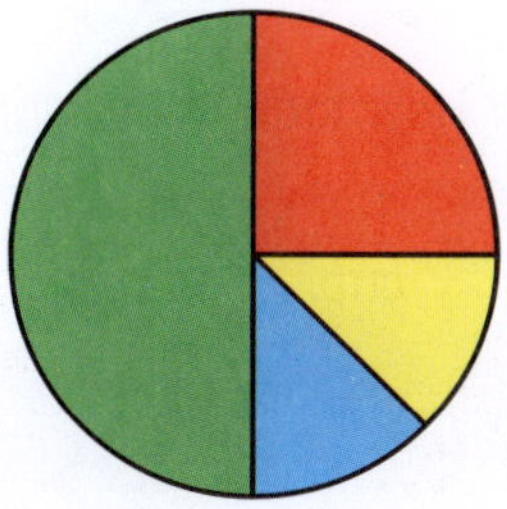

⑦ What fraction of the class voted for red? ______

⑧ What fraction of the class voted for green? _______

What are the unknowns in these equivalent number sentences?

⑨ 27 + ___ = 37 + 25

⑩ 56 − ___ = 76 − 36

⑪ 59 − 24 = 16 + ___

⑫ 68 − 43 = ___ − 15

⑬ **26 55 63**
Which two of these numbers will add to an even answer? ___ and ___

⑭ **26 35 27**
Which two of these numbers will add to an odd answer? ___ and ___

⑮ Draw a right-angle triangle in the box.
Make side AB the base.
Draw a 4 cm line vertically from A to C.
Join B to C.
Label each side with its length.

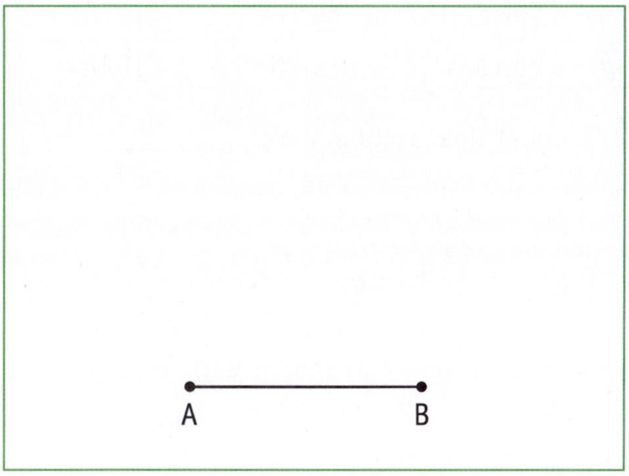

SCORE /15 0-7 8-13 14-15

Review: Unit 29 and Unit 30

AC9M4N03, AC9M4N06, AC9M4N08, AC9M4M03, AC9M4P01, AC9M4SP03, AC9M4ST03

What is the safest route?

(1) From 3·6 to 40

3·6 + ______ = ______

______ × ______ = 40

(2) From 6·7 to 700

______ + ______ = ______

______ × ______ = ______

(3) Chef has 6 identical pizzas to make. How many of each item should he put on each pizza?

- 12 olives — Put ___ olives on each pizza.
- 3 pineapple slices — Put ___ a slice on each pizza.
- 18 slices of salami — Put ___ salami slices on each pizza.

(4) What is the chance of pulling a red counter out of this bag?

- ◯ Impossible
- ◯ Unlikely
- ◯ Likely
- ◯ Very likely
- ◯ Certain

(5) How many hours between 9:30 pm and 2:30 am? __________ hours

(6) $\frac{1}{2}$ cup + $\frac{1}{4}$ cup = ___ cup

(7) $\frac{1}{4}$ cup + $\frac{1}{8}$ cup + $\frac{1}{8}$ cup = ___ cup

(8) Half of $35 = $________

(9) Quarter of $11 = $________

Use this spreadsheet to answer the following questions.

	A	B	C	D
1	**Packet of**	**OJ**	**Big Orange**	**C-vit Orange**
2	1 carton	$3·45	$4·20	$3·85
3	2 cartons	$6·45	$8·00	$7·40
4	3 cartons	$9·90	$11·70	$10·80
5	4 cartons	$13·00	$15·50	$14·40

(10) What amount does cell D2 show? _________

(11) How much is 4 cartons of Big Orange? _________

(12) How much more is 2 cartons of C-vit Orange than 2 cartons of OJ? ________

(13) Use the grid to show two shapes with an area of 8 squares.

Each shape should have at least one line of symmetry.

Mark the lines of symmetry.

SCORE /13 (0-6) (7-11) (12-13)

TARGETING MATHS HOMEWORK: YEAR 4 © PASCAL PRESS ISBN: 9781925726572

Review: Unit 31 and Unit 32

AC9M4N07, AC9M4N08, AC9M4M02, AC9M4SP01, AC9M4ST01

Use the fraction lines to compare the fractions. Then circle the larger fraction.

① $\frac{5}{6}$ or $\frac{5}{8}$

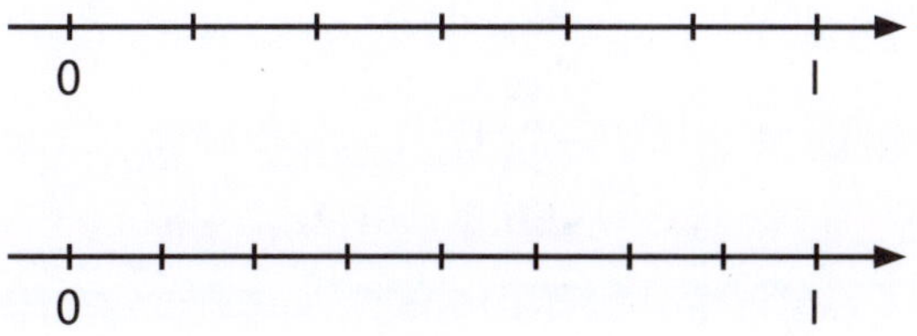

② $\frac{3}{5}$ or $\frac{3}{8}$

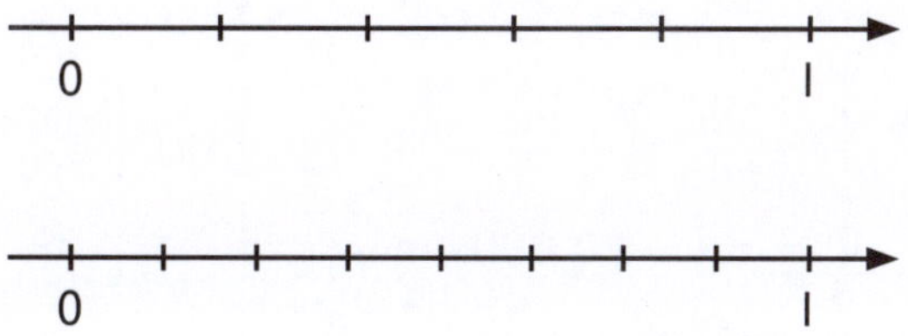

③ Find $\frac{1}{4}$ of these counters and complete the number sentences.

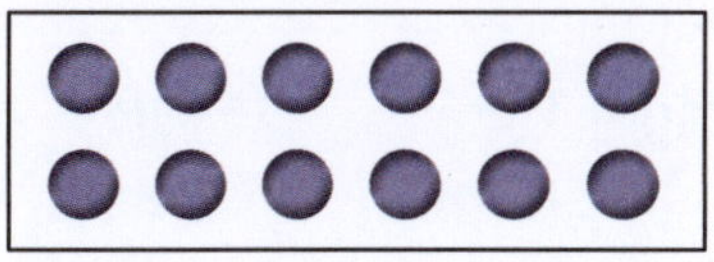

— of ☐ = ☐ and ☐ ÷ ☐ = ☐

There are 24 students in Mrs Green's class and the pie graph shows their Spellathon results.

④ Which score out of 100 was the most common?

____

⑤ Which two scores had the same amount?

__________ and __________

⑥ What fraction of the students scored 81 to 90?

_____

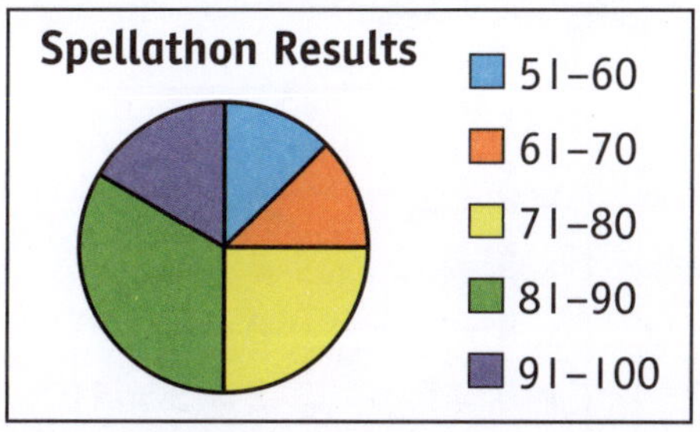

Use the grid to help you answer these questions.

⑦ How much fence is needed for a garden bed that is 4 metres by 6 metres?

____

⑧ What is the area of this garden bed? _____

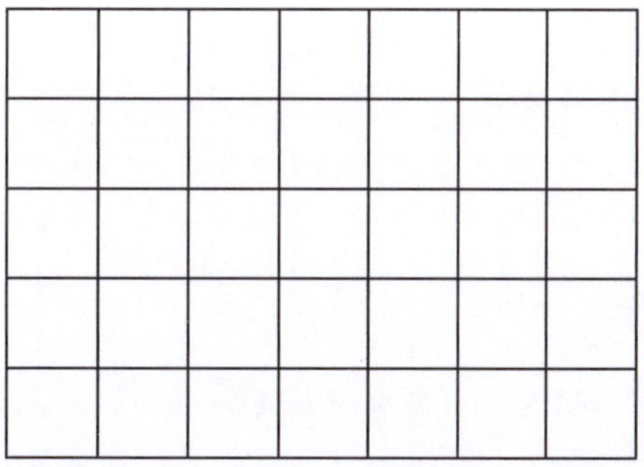

What is the value?

⑨ Fifty 5-cent coins ________

⑩ Thirteen 20-cent coins ________

⑪ Thirty-one 50-cent coins ________

⑫ One hundred and one 10-cent coins ________

⑬ What is the area of this shape?

______ squares

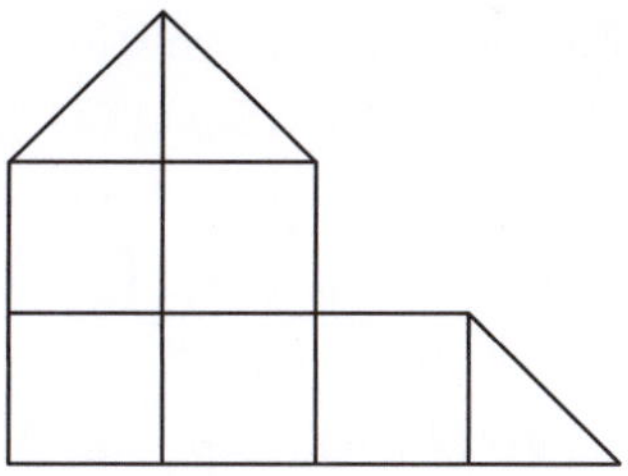

SCORE /13 0-6 7-11 12-13

★ QUICK MATHS FACTS ★

Measurement

Length

1 km = 1000 m

1 m = 100 cm

1 cm = 10 mm

Area

1 m^2 = 10 000 cm^2

1 hectare = 10 000 m^2

1 km^2 = 1 000 000 m^2

Capacity

1 L = 1000 mL

Mass

1 t = 1000 kg

1 kg = 1000 g

Fractions, decimals and percentages

$\frac{1}{100}$ = 0·01 = 1%

$\frac{25}{100}$ = 0·25 = 25%

$\frac{50}{100}$ = 0·5 = 50%

$\frac{75}{100}$ = 0·75 = 75%

$\frac{100}{100}$ = 1 = 100%

Place value

Ones	1
Tens	12
Hundreds	123
Thousands	1234
Ten thousands	12 345
Hundred thousands	123 456
Millions	1 234 567
Ten millions	12 345 678
Hundred millions	123 456 789

Time

60 s = 1 min

60 min = 1 h

1440 min = 1 day

24 h = 1 day

7 days = 1 week

365 days = 1 year

52 weeks = 1 year

12 months = 1 year

Multiplication tables

1 × 1 = 1	2 × 1 = 2	3 × 1 = 3	4 × 1 = 4	5 × 1 = 5	6 × 1 = 6
1 × 2 = 2	2 × 2 = 4	3 × 2 = 6	4 × 2 = 8	5 × 2 = 10	6 × 2 = 12
1 × 3 = 3	2 × 3 = 6	3 × 3 = 9	4 × 3 = 12	5 × 3 = 15	6 × 3 = 18
1 × 4 = 4	2 × 4 = 8	3 × 4 = 12	4 × 4 = 16	5 × 4 = 20	6 × 4 = 24
1 × 5 = 5	2 × 5 = 10	3 × 5 = 15	4 × 5 = 20	5 × 5 = 25	6 × 5 = 30
1 × 6 = 6	2 × 6 = 12	3 × 6 = 18	4 × 6 = 24	5 × 6 = 30	6 × 6 = 36
1 × 7 = 7	2 × 7 = 14	3 × 7 = 21	4 × 7 = 28	5 × 7 = 35	6 × 7 = 42
1 × 8 = 8	2 × 8 = 16	3 × 8 = 24	4 × 8 = 32	5 × 8 = 40	6 × 8 = 48
1 × 9 = 9	2 × 9 = 18	3 × 9 = 27	4 × 9 = 36	5 × 9 = 45	6 × 9 = 54
1 × 10 = 10	2 × 10 = 20	3 × 10 = 30	4 × 10 = 40	5 × 10 = 50	6 × 10 = 60
1 × 11 = 11	2 × 11 = 22	3 × 11 = 33	4 × 11 = 44	5 × 11 = 55	6 × 11 = 66
1 × 12 = 12	2 × 12 = 24	3 × 12 = 36	4 × 12 = 48	5 × 12 = 60	6 × 12 = 72

7 × 1 = 7	8 × 1 = 8	9 × 1 = 9	10 × 1 = 10	11 × 1 = 11	12 × 1 = 12
7 × 2 = 14	8 × 2 = 16	9 × 2 = 18	10 × 2 = 20	11 × 2 = 22	12 × 2 = 24
7 × 3 = 21	8 × 3 = 24	9 × 3 = 27	10 × 3 = 30	11 × 3 = 33	12 × 3 = 36
7 × 4 = 28	8 × 4 = 32	9 × 4 = 36	10 × 4 = 40	11 × 4 = 44	12 × 4 = 48
7 × 5 = 35	8 × 5 = 40	9 × 5 = 45	10 × 5 = 50	11 × 5 = 55	12 × 5 = 60
7 × 6 = 42	8 × 6 = 48	9 × 6 = 54	10 × 6 = 60	11 × 6 = 66	12 × 6 = 72
7 × 7 = 49	8 × 7 = 56	9 × 7 = 63	10 × 7 = 70	11 × 7 = 77	12 × 7 = 84
7 × 8 = 56	8 × 8 = 64	9 × 8 = 72	10 × 8 = 80	11 × 8 = 88	12 × 8 = 96
7 × 9 = 63	8 × 9 = 72	9 × 9 = 81	10 × 9 = 90	11 × 9 = 99	12 × 9 = 108
7 × 10 = 70	8 × 10 = 80	9 × 10 = 90	10 × 10 = 100	11 × 10 = 110	12 × 10 = 120
7 × 11 = 77	8 × 11 = 88	9 × 11 = 99	10 × 11 = 110	11 × 11 = 121	12 × 11 = 132
7 × 12 = 84	8 × 12 = 96	9 × 12 = 108	10 × 12 = 120	11 × 12 = 132	12 × 12 = 144

Thirty days have September, April, June and November.
All the rest have thirty-one, except February alone...
which has twenty-eight days clear, and twenty-nine each leap year.

Answers - Homework Year 4

TERM 1

UNIT 1 NUMBER & ALGEBRA

Introducing Decimals

1 $\frac{1}{10}$ or 0·1
2 $\frac{3}{10}$ or 0·3
3 $\frac{5}{10}$ or 0·5
4 $\frac{9}{10}$ or 0·9
5 $\frac{7}{10}$ or 0·7
6 $\frac{4}{10}$ or 0·4
7 $\frac{12}{10}$ or 1·2
8 $\frac{22}{10}$ or 2·2
9 $\frac{18}{10}$ or 1·8
10 $\frac{14}{10}$ or 1·4

Quick Check

1 0·4
2 $\frac{7}{10}$
3 0·5
4 0·3
5 $\frac{9}{10}$
6 1·4
7 $\frac{13}{10}$
8 1·7
9 $\frac{21}{10}$
10 2·8

Odd and Even Numbers

1 odd
2 even
3 14, even
4 7, 8, 9, 10, 11, 12, 13, 14, 15, 16.
5 even
6 odd
7 odd
8 even
9 even
10 even

UNIT 1 STATISTICS & PROBABILITY

Birthdays by the Seasons

1 56
2 80
3 69
4 56
5 Zoe
6 Only 56 in summer which is the lowest.

UNIT 1 MEASUREMENT & SPACE

Joining Shapes

1

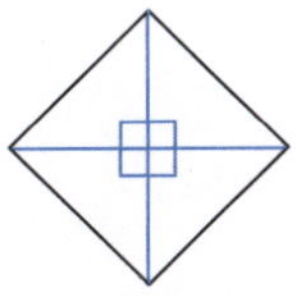

2

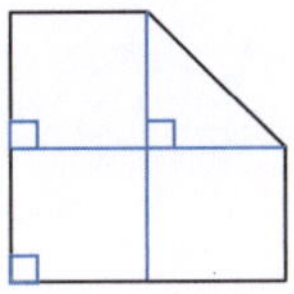

3

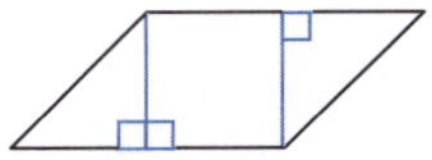

4

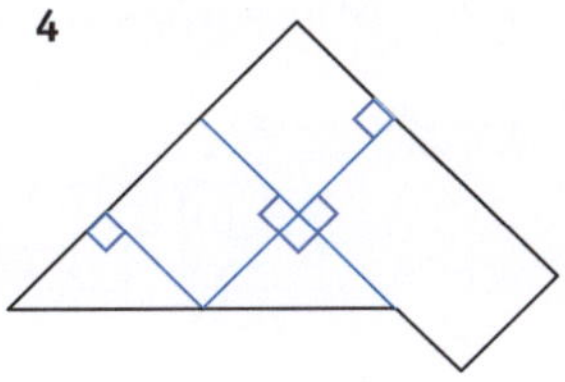

UNIT 1 NAPLAN*-STYLE QUESTIONS

1

2 41
3 24

UNIT 1 PROBLEM OF THE WEEK

The Greengrocer's Apple Graph

1 Tuesday
2 Thursday

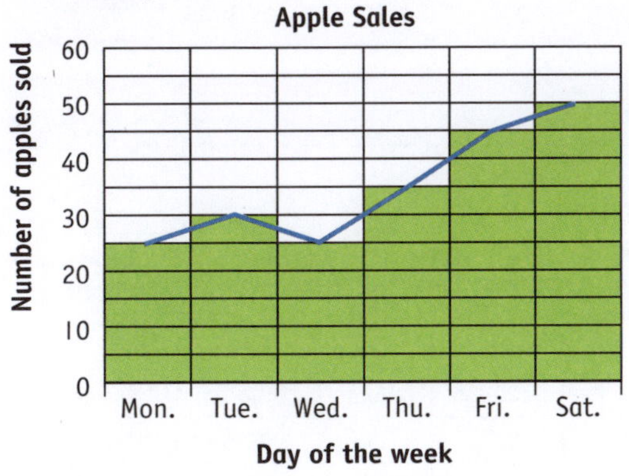

3 No, because Saturday is now the tallest.
4 Yes, because you do not have to count the 5s and 10s.

UNIT 2 NUMBER & ALGEBRA

Equivalent Fractions

1 $\frac{1}{2}$
2 $\frac{1}{4}, \frac{2}{4}, \frac{3}{4}$
3 $\frac{1}{8}, \frac{2}{8}, \frac{3}{8}, \frac{4}{8}, \frac{5}{8}, \frac{6}{8}, \frac{7}{8}$
4 $\frac{2}{8}$
5 $\frac{1}{2}$ and $\frac{2}{4}$
6 $\frac{3}{4}$
7 1
8 $\frac{2}{3}$
9 $\frac{1}{3}, \frac{1}{2}, \frac{2}{3}, \frac{5}{6}$
10 $\frac{3}{5}$
11 $\frac{5}{10}, \frac{9}{10}$
12 $\frac{3}{6}$
13 $\frac{5}{10}$

Quick Check

1 $\frac{2}{8}$
2 $\frac{1}{2}$
3 $\frac{1}{2}$
4 $\frac{2}{10}$
5 $\frac{2}{6}$
6 $\frac{6}{10}$
7 $\frac{2}{2}$ or 1
8 $\frac{2}{3}$
9 $\frac{3}{5}$
10 $\frac{6}{8}$

Multiplying by 10 and 100

1 230
2 1060
3 360
4 1270
5 1450
6 3500
7 3600
8 14 500
9 4500
10 36 000

UNIT 2 STATISTICS & PROBABILITY

Just by Chance

1 Chen
2 Chen
3 3 or close
4 2
5 Likely

UNIT 2 MEASUREMENT & SPACE

Telling the Time

1 4:00
2 3:30
3 9:15
4 5:55
5 5:43
6 6:50
7 500 mL
8 250 mL
9 1500 mL

UNIT 2 NAPLAN*-STYLE QUESTIONS

1 $\frac{2}{6}$
2 $117
3 The third spinner

UNIT 2 PROBLEM OF THE WEEK

A Scissors-Paper-Rock Investigation

Answers will vary.

UNIT 3 NUMBER & ALGEBRA

Rounding and Estimating

1 30
2 40
3 30
4 70
5 100
6 300
7 400
8 400
9 down, 370, 10
up, 400, 100
10 up, 420, 10
down, 400, 100

Quick Check

1 60
2 30
3 100
4 150
5 200
6 600
7 400
8 400
9 900
10 300

Addition And Subtraction with Unknowns

1 10, 8, 2
2 12, 4, 8
3 9, 3, 6

ANSWERS

UNIT 3 STATISTICS & PROBABILITY

Cordial Survey

1, 2, 4 See graph
3 6

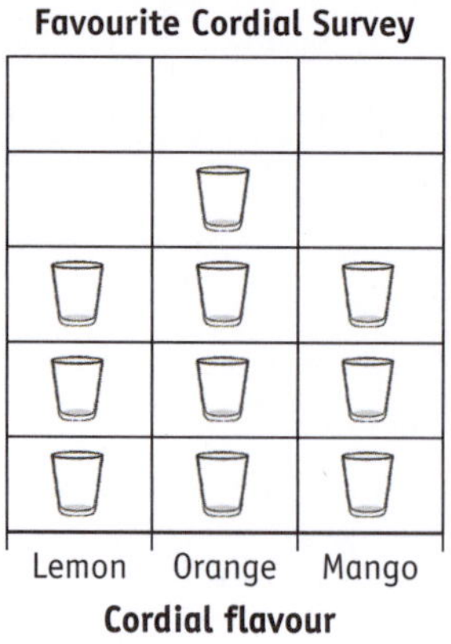

UNIT 3 MEASUREMENT & SPACE

Perimeter

Estimates should be within 2 of given answer.

1 Estimate 8, Check 12
2 Estimate 6, Check 10
3 Estimate 10, Check 14
4 Estimate 10, Check 14
5 Estimate 14, Check 12
6 Estimate 9, Check 12

UNIT 3 NAPLAN*-STYLE QUESTIONS

1 245 2 16 3 100 cm

UNIT 3 PROBLEM OF THE WEEK

Perimeter Puzzles

1 2 × 4 + 2 × 3 + 2 × 1
= 8 + 6 + 2 = 18
2 3 × 2 + 4 × 1 + 4
= 6 + 4 + 4 = 14
3 3 + 4 × 1 + 5
= 3 + 4 + 5 = 12
4 2 × 2 + 2 × 1 + 2 × 3
= 4 + 2 + 6 = 12

Dan's challenge

The shortest perimeter is 12.
2 × 2 + 4 × 2 = 4 + 8 = 12

UNIT 4 NUMBER & ALGEBRA

Estimation and Addition

1 621
Estimate 1: 300 + 300 = 600
Estimate 2: 280 + 340 = 620
2 1208
Estimate 1: 600 + 600 = 1200
Estimate 2: 570 + 640 = 1210
3 899
Estimate 1: 500 + 400 = 900
Estimate 2: 470 + 430 = 900
4 865
Estimate 1: 500 + 400 = 900
Estimate 2: 490 + 380 = 870
5 772
Estimate 1: 300 + 400 = 700
Estimate 2: 320 + 450 = 770
6 805
Estimate 1: 300 + 500 = 800
Estimate 2: 350 + 460 = 810

Quick Check

1 340 + 890 = 1230
2 280 + 250 = 530
3 390 + 430 = 820
4 490 + 230 = 720
5 380 + 230 = 610
6 900 + 400 = 1300
7 700 + 600 = 1300
8 700 + 500 = 1200
9 900 + 900 = 1800
10 1300 + 2500 = 3800

Calculator Sequences

1 126 + 56 = 182
2 96 − 18 = 78
3 48 − 13 = 35
4 28 + 7 = 35
5 32 + 129 = 161
6 24 − 15 = 9 (still too many!)

UNIT 4 STATISTICS & PROBABILITY

At the Birdbath

1 90 2 30 3 15
4

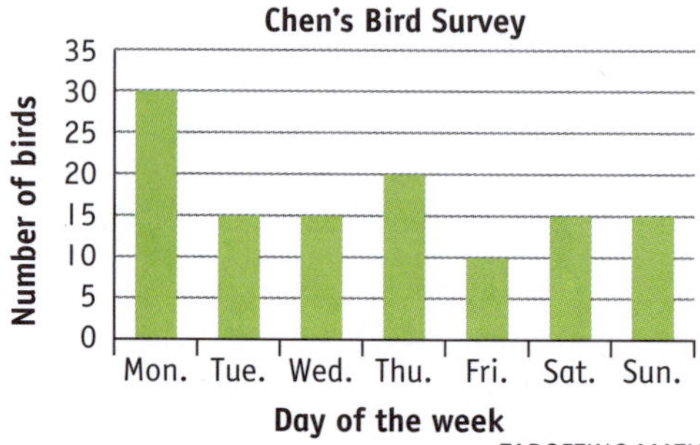

UNIT 4 MEASUREMENT & SPACE

Grid Coordinates

1 A3
2 F3
3 D1
4 C4
5 F2
6 B5
7 E4
8 G5
9 B2
10 D3

UNIT 4 NAPLAN*-STYLE QUESTIONS

1 $44 2 800 3 C3

UNIT 4 PROBLEM OF THE WEEK

All but Q1 and Q4 are open ended.

1 Totals are 208 and 210
4 Totals are 1588 and 1600

UNIT 5 NUMBER & ALGEBRA

Working with Tenths

1 1·5
2 1·8
3 1·3
4 1·4
5 1·7

Quick Check

1 0·8 or $\frac{8}{10}$
2 1·3 or $\frac{13}{10}$
3 1·0 or $\frac{10}{10}$
4 1·1 or $\frac{11}{10}$
5 1·1 or $\frac{11}{10}$
6 1·5 or $\frac{15}{10}$
7 1·6 or $\frac{16}{10}$
8 2·4 or $\frac{24}{10}$
9 1·8 or $\frac{18}{10}$
10 1·9 or $\frac{19}{10}$

Estimation and Subtraction

1 Estimate: 330
Actual: 325
2 Estimate: 140
Actual: 142
3 Estimate: 520
Actual: 523
4 Estimate: 210
Actual: 195
5 Estimate: 580
Actual: 588
6 Estimate: 130
Actual: 135

UNIT 5 STATISTICS & PROBABILITY

Counting Colours

Open ended

UNIT 5 MEASUREMENT & SPACE

am and pm Times

1–5

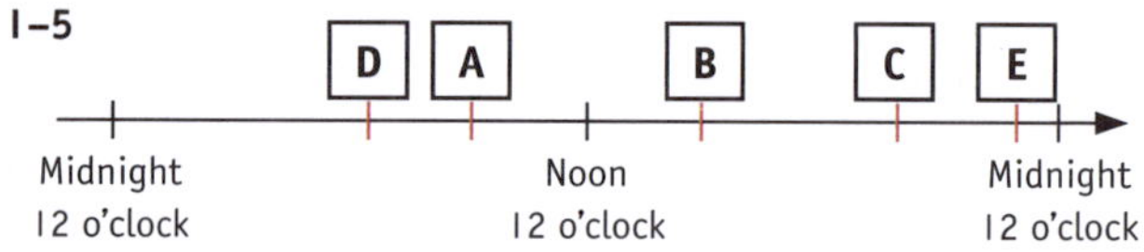

6 B 7 E 8 C 9 D 10 A

UNIT 5 NAPLAN*-STYLE QUESTIONS

1 90 2 likely 3 1:00 am

UNIT 5 PROBLEM OF THE WEEK

A Visit to Fantasy World

Jo's timeline

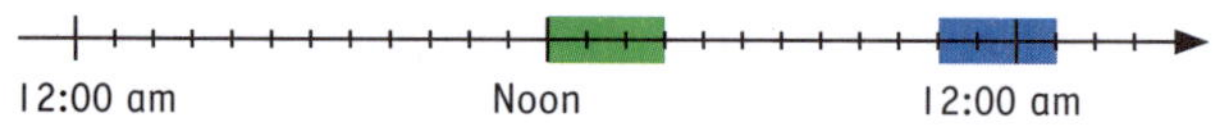

Zoe's timeline

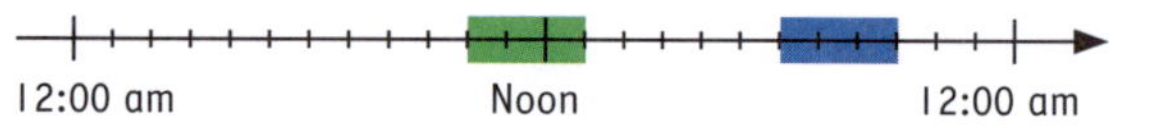

Kai's timeline

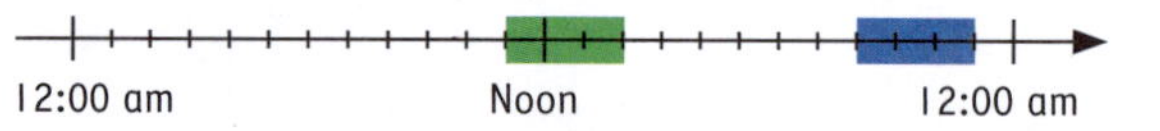

	Time left town	At Fantasy World			Time arrived back in town
		Arrive	Time spent	Leave	
Jo	12:00 pm	3:00 pm	7 hours	10:00 pm	1:00 am
Zoe	10:00 am	1:00 pm	5 hours	6:00 pm	9:00 pm
Kai	11:00 am	2:00 pm	6 hours	8:00 pm	11:00 pm

5 Yes

TARGETING MATHS HOMEWORK: YEAR 4 © PASCAL PRESS ISBN: 9781925726572

UNIT 6 NUMBER & ALGEBRA

Multiplying Money Amounts

1 2	3 10	5 $3·60	7 $3·25	9 $1·80
2 5	4 20	6 $2·10	8 $3·00	10 $2·80

Quick Check

1 $3	3 90c	5 50c	7 $1·50	9 $1·40
2 $1·40	4 $1	6 $1·20	8 $1·20	10 $2

Multiplication Facts

1 $6 \times 6 = 6 \times 5 + 6 \times 1 = 30 + 6 = 36$
2 $6 \times 7 = 6 \times 5 + 6 \times 2 = 30 + 12 = 42$
3 $8 \times 6 = 8 \times 5 + 8 \times 1 = 40 + 8 = 48$
4 $9 \times 7 = 9 \times 5 + 9 \times 2 = 45 + 18 = 63$
5 $7 \times 8 = 7 \times 5 + 7 \times 3 = 35 + 21 = 56$

UNIT 6 STATISTICS & PROBABILITY

Midday and Midnight Temperatures

1 5 °C	3 7 °C	5 Wednesday
2 12 °C	4 Friday	

UNIT 6 MEASUREMENT & SPACE

Lines of Symmetry

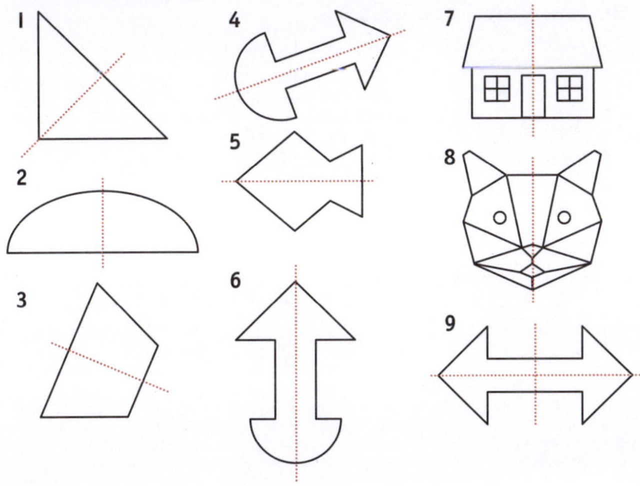

UNIT 6 NAPLAN*-STYLE QUESTIONS

1 $8　　2 36 apples　　3 12

UNIT 6 PROBLEM OF THE WEEK

With Seven Squares

Open ended

UNIT 7 NUMBER & ALGEBRA

Odd and Even Numbers

1 One odd and one even, all odd totals
2 Two odd or two even, all even totals
3 Open ended, but totals of 2, 4, 6, 8, 10 are possible.
4 Open ended, but totals of 4, 6, 8, 10, 12 are possible.
5 Open ended, but totals of 3, 5, 7, 9, 11 are possible.

Quick Check

1 Even	3 Even	5 Even	7 Even	9 Even
2 Even	4 Even	6 Even	8 Even	10 Odd

Number Sequences

	Sequence	Rule	Odd or Even
1	1, 4, 3, 6, 5, 8, 7, 10, ...	+ 3, − 1	Odd
2	2, 4, 3, 5, 4, 6, 5, 7 ...	+2, − 1	Odd
3	3, 5, 4, 6, 5, 7, 6, 9 ...	+ 2, − 1	Even
4	2, 8, 14, 20, 26, 32, 38, 44 ...	+ 6	Even
5	3, 8, 13, 18, 23, 28, 33, 38 ...	+ 5	Odd
6	4, 8, 5, 9, 6, 10, 7, 11 ...	+ 4, − 3	Odd
7	13, 12, 14, 13, 15, 14, 16, 15 ...	− 1, + 2	Even
8	5, 15, 13, 23, 21, 31, 29, 39 ...	+ 10, − 2	Odd
9	2, 4, 8, 16, 32, 64, 128, 256 ...	double	Even
10	32, 41, 39, 48, 46, 54, 52, 61 ...	+ 9, − 2	Even

UNIT 7 STATISTICS & PROBABILITY

Yesterday's Weather

1 16 °C
2 Between 6:00 am and 12:00 pm
3 Between 12:00 pm and 6:00 pm
4 The line graph
5 Open ended

UNIT 7 MEASUREMENT & SPACE

All about Angles

1 right	4 obtuse	7 A right
2 acute	5 reflex	8 B acute
3 straight	6 C obtuse	9 D reflex

UNIT 7 NAPLAN*-STYLE QUESTIONS

1 543

2

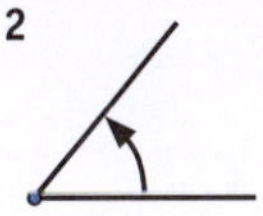

3 35 tonnes

UNIT 7 PROBLEM OF THE WEEK

Crossing Lines

Open ended

UNIT 8 NUMBER & ALGEBRA

Finding Doubles, Finding Halves

1 126	3 78	5 33	7 47
2 152	4 326	6 42	8 84

Quick Check

1 56	4 116	7 15	10 35	13 29
2 72	5 246	8 18	11 45	14 56
3 158	6 12	9 25	12 37	15 62

Multiplication

1 $10 \times 26 = 260$, $5 \times 26 = 130$
2 $10 \times 44 = 440$, $5 \times 44 = 220$
3 $10 \times 38 = 380$, $5 \times 38 = 190$
4 $10 \times 62 = 620$, $5 \times 62 = 310$
5 $10 \times 96 = 960$, $5 \times 96 = 480$
6 $10 \times 64 = 640$, $5 \times 64 = 320$
7 $10 \times 76 = 760$, $5 \times 76 = 380$
8 $10 \times 54 = 540$, $5 \times 54 = 270$

UNIT 8 MEASUREMENT & SPACE

Scale it Up

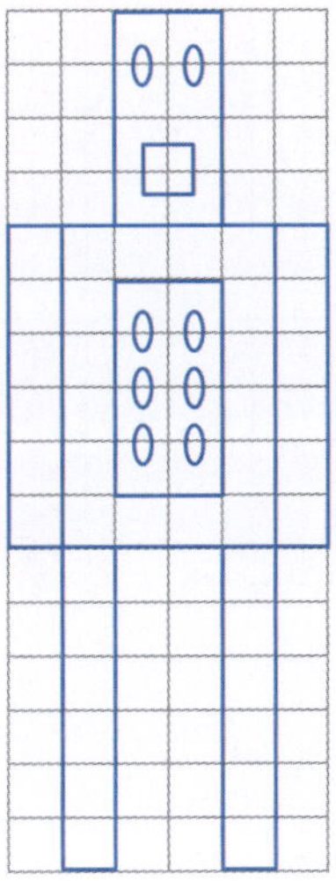

ANSWERS

UNIT 8 STATISTICS & PROBABILITY

North, East, South or West?

1 Yes
2 Yes. All 4 quadrants have an equal chance.
3 NE: 2, SE: 2, SW: 2, NW: 2
4 Open ended
5 Open ended

UNIT 8 NAPLAN*-STYLE QUESTIONS

1 25
2 $175
3 Likely

UNIT 8 PROBLEM OF THE WEEK

Heads or Tails

1 The possibilities are H H H, T H H, T T H and T T T. Note: T H H and H T H are considered the same as the order in which the coins are tossed is not determined.
2 Open ended
3 Open ended
4 Jake and Kai made the best predictions as there is the same chance of getting either H T T or T H H.

TERM 1 REVIEW

Review: Unit 1 and Unit 2

1 0·6
2 0·3
3 1·2
4 $\frac{7}{10}$
5 $\frac{4}{10}$
6 $\frac{16}{10}$
7 12, 32, 42, 50
8 2nd and 3rd shapes
9 $\frac{1}{4}$, $\frac{3}{4}$
10 $\frac{1}{8}$, $\frac{2}{8}$, $\frac{3}{8}$, $\frac{4}{8}$
11 $\frac{6}{8}$
12 320
13 90
14 700
15 1250
16 1900
17 4000
18 500
19 7500
20 4000
21 82 000
22 1st spinner

Review: Unit 3 and Unit 4

1 50
2 40
3 50
4 100
5 100
6 300
7 300
8 1000
9 6 with 4 and 2
10 10 with 2 and 8
11 10 cm
12 12 cm
13 16 cm
14 Key: 5 people
15 880, 800, 886
16 780, 800, 779
17 850, 900, 848
18 1200, 1200, 1191
19 $196 + $73 = $269

Review: Unit 5 and Unit 6

1 Show 1·4 as:

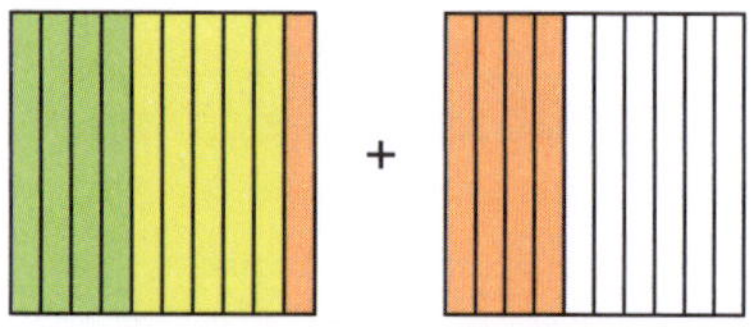

2 Show 1·5 as:

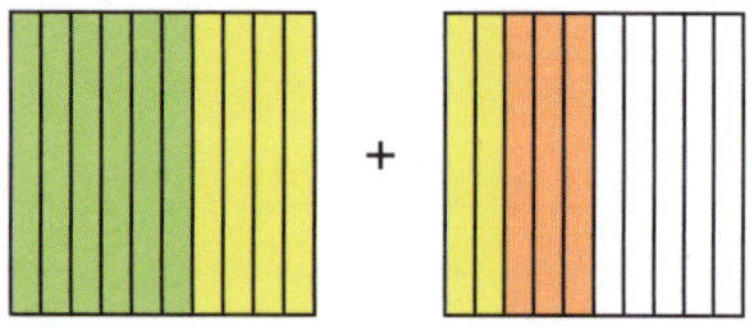

3 Estimate: 550 + 330 = 220
Actual: 219
4 Estimate: 1370 − 900 = 470
Actual: 474
5

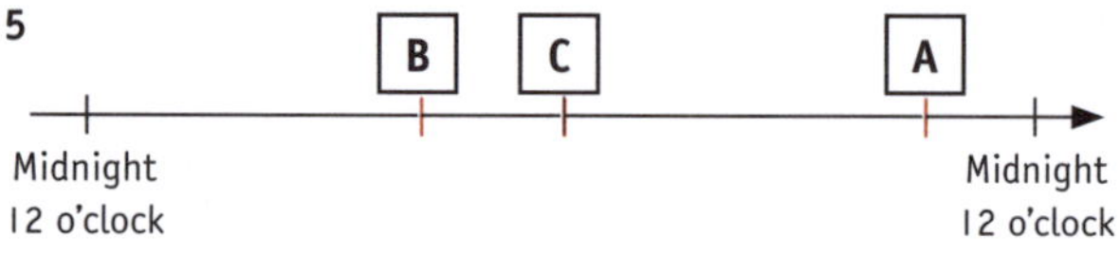

6

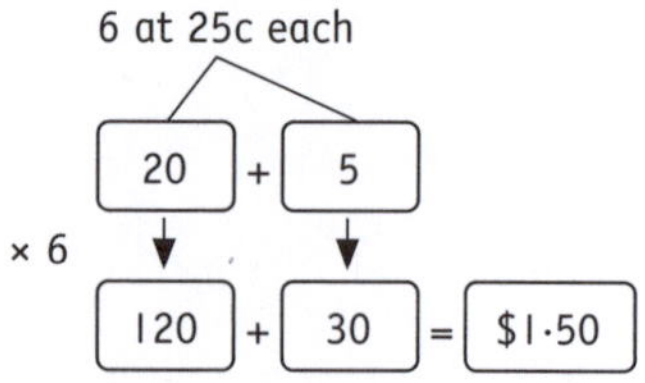

7

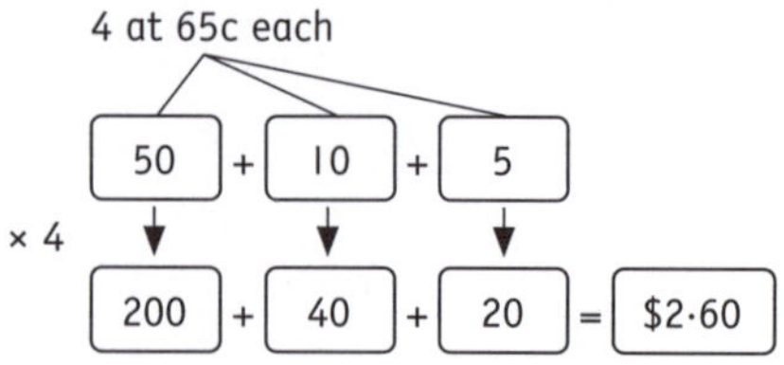

8 6 × 8 = 6 × 5 + 6 × 3 = 30 + 18 = 48
9 7 × 6 = 7 × 5 + 7 × 1 = 35 + 7 = 42
10 8 × 9 = 8 × 5 + 8 × 4 = 40 + 32 = 72
11

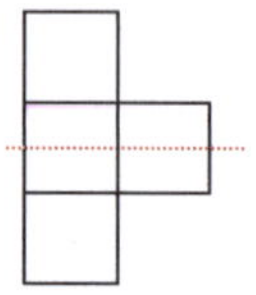

12

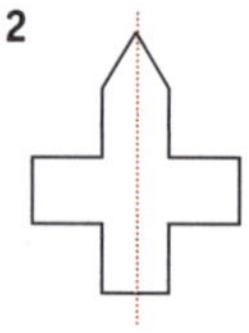

13

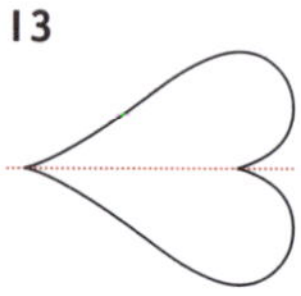

Review: Unit 7 and Unit 8

1 odd
2 odd
3 even
4 even
5 even
6 even
7 6, 9, 7 with the rule + 3, − 2
8 32, 37, 42 with the rule + 5
9 51, 52, 49 with the rule − 3, + 1
10 9 degrees
11 14 degrees
12 2 degrees
13 right
14 acute
15 obtuse
16 72
17 98
18 218
19 14
20 27
21 104
22 10 × 32 = 320 and 5 × 32 = 160
23 10 × 46 = 460 and 5 × 46 = 230

TERM 2

UNIT 9 NUMBER & ALGEBRA

Counting in Tenths

1	$\frac{7}{10}$	$\frac{8}{10}$	$\frac{9}{10}$	1	$1\frac{1}{10}$	$1\frac{2}{10}$	$1\frac{3}{10}$	$1\frac{4}{10}$	$1\frac{5}{10}$
	0·7	0·8	0·9	1	1·1	1·2	1·3	1·4	1·5
2	$\frac{15}{10}$	$\frac{16}{10}$	$\frac{17}{10}$	$\frac{18}{10}$	$\frac{19}{10}$	$\frac{20}{10}$	$\frac{21}{10}$	$\frac{22}{10}$	$\frac{23}{10}$
	1·5	1·6	1·7	1·8	1·9	2	2·1	2·2	2·3
3	$\frac{5}{10}$	$\frac{7}{10}$	$\frac{9}{10}$	$\frac{11}{10}$	$\frac{13}{10}$	$\frac{15}{10}$	$\frac{17}{10}$	$\frac{19}{10}$	$\frac{21}{10}$
	0·5	0·7	0·9	1·1	1·3	1·5	1·7	1·9	2·1

4 0·9
5 1·1
6 1·8
7 0·8
8 0·5
9 1·9
10 0·7

Quick Check

1 0·8
2 1·3
3 0·7
4 1·1
5 2·2
6 0·6
7 0·9
8 0·5
9 1·8
10 1·2

All about Fractions

1 $\frac{2}{4}$, $\frac{3}{4}$, 1, $1\frac{1}{4}$, $1\frac{2}{4}$, $1\frac{3}{4}$, 2, $2\frac{1}{4}$, $2\frac{2}{4}$, $2\frac{3}{4}$, 3
2 $\frac{1}{3}$, $\frac{2}{3}$, 1, $1\frac{1}{3}$, $1\frac{2}{3}$, 2, $2\frac{1}{3}$, $2\frac{2}{3}$, 3, $3\frac{1}{3}$, $3\frac{2}{3}$, 4
3 $\frac{3}{5}$, $\frac{4}{5}$, 1, $1\frac{1}{5}$, $1\frac{2}{5}$, $1\frac{3}{5}$, $1\frac{4}{5}$, 2, $2\frac{1}{5}$, $2\frac{2}{5}$
4 $\frac{2}{4}$, $\frac{3}{4}$, $\frac{4}{4}$, $\frac{5}{4}$, $\frac{6}{4}$, $\frac{7}{4}$, $\frac{8}{4}$, $\frac{9}{4}$, $\frac{10}{4}$, $\frac{11}{4}$
5 $\frac{1}{3}$, $\frac{2}{3}$, $\frac{3}{3}$, $\frac{4}{3}$, $\frac{5}{3}$, $\frac{6}{3}$, $\frac{7}{3}$, $\frac{8}{3}$, $\frac{9}{3}$, $\frac{10}{3}$
6 $\frac{3}{5}$, $\frac{4}{5}$, $\frac{5}{5}$, $\frac{6}{5}$, $\frac{7}{5}$, $\frac{8}{5}$, $\frac{9}{5}$, $\frac{10}{5}$, $\frac{11}{5}$, $\frac{12}{5}$

TARGETING MATHS HOMEWORK: YEAR 4 © PASCAL PRESS ISBN: 9781925726572

UNIT 9 STATISTICS & PROBABILITY

Fast Food Survey

1

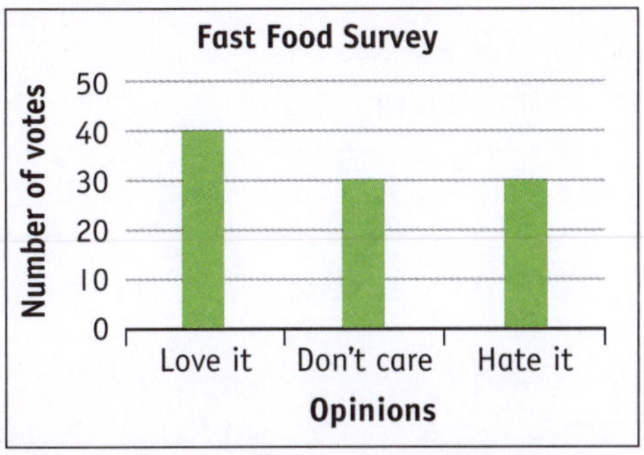

2 40 3 30 4 30 5 20

UNIT 9 MEASUREMENT & SPACE

Measuring in Centimetres or Millimetres

1 4 cm or 40 mm
2 $3\frac{1}{2}$ cm or 35 mm
3 5 cm or 50 mm
4 $5\frac{1}{2}$ cm or 55 mm
5 $6\frac{1}{2}$ cm or 65 mm
6 125 mm
7 1250 mm
8 11 cm or 110 mm

UNIT 9 NAPLAN*-STYLE QUESTIONS

1 3·1 2 The highlighter 3 10 m

UNIT 9 PROBLEM OF THE WEEK

Don't Cross the Line

Push totals: Claire 395 mm, Chen 435 mm, Jo 460 mm, Zoe 441 mm

1 Longest single push was Claire Push 2: 15 cm.
2 Shortest single push was Zoe Push 3: 55 mm.
3 Closest: Jo
4 Furthest: Claire
5 Difference between Jo and Claire: 65 mm

UNIT 10 NUMBER & ALGEBRA

Introducing Hundredths

1 $\frac{25}{100}$
2 $\frac{36}{100}$
3 $\frac{58}{100}$
4 $\frac{3}{10}$ or $\frac{30}{100}$
5 $\frac{7}{10}$ or $\frac{70}{100}$
6 $\frac{6}{10}$ or $\frac{60}{100}$
7 $\frac{9}{10}$ or $\frac{90}{100}$
8 $\frac{8}{10}$ or $\frac{80}{100}$
9 $\frac{10}{10}$ or $\frac{100}{100}$

Quick Check

1 thirty
2 fifty
3 seventy
4 ten
5 one hundred and ten
6 five
7 eight
8 one
9 ten
10 twelve

The Area Model for Multiplication

1

×	20	+ 6	
6	120	36	= 156

2

×	20	+ 7	
3	60	21	= 81

3

×	30	+ 8	
4	120	32	= 152

4

×	70	+ 2	
3	210	6	= 216

5

×	100	+ 20	+ 4	
4	400	80	16	= 496

6

×	200	+ 50	+ 3	
6	1200	300	18	= 1518

UNIT 10 STATISTICS & PROBABILITY

Take Care

1 Preferred peas 10, potatoes 16, cabbage 8 or pumpkin 14.
2 48
3

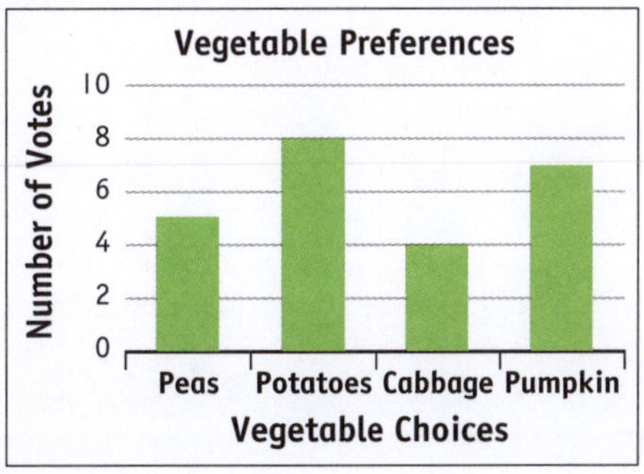

4 Preferred peas 5, potatoes 8, cabbage 4 or pumpkin 7.
5 24

UNIT 10 MEASUREMENT & SPACE

Reading Temperature

1 A - 30 °C
2 B - 20 °C
3 C - 40 °C
4 D - 22 °C
5 E - 36 °C
6 3
7 2

UNIT 10 NAPLAN*-STYLE QUESTIONS

1 1·15
2 6 cm
3 6 °C

UNIT 10 PROBLEM OF THE WEEK

Invent a Rule

1 128
2 68
3 To make the largest product, arrange the numbers in order, highest - middle - lowest. Multiply the highest number by the 2-digit number made by the middle and lowest.
To make the smallest product, reverse the order of the digits and proceed as before.

4–6 Open ended

UNIT 11 NUMBER & ALGEBRA

Decimal Notation

1 1·45 or $1\frac{45}{100}$
2 0·85 or $\frac{85}{100}$
3 2·16 or $2\frac{16}{100}$
4 1·36 or $1\frac{36}{100}$
5 1·77 or $1\frac{77}{100}$
6 1·9 or $1\frac{9}{10}$

Quick Check

1 $1\frac{3}{10}$ or $\frac{13}{10}$
2 $1\frac{35}{100}$ or $\frac{135}{100}$
3 $1\frac{3}{100}$ or $\frac{103}{100}$
4 $\frac{76}{100}$
5 $2\frac{38}{100}$ or $\frac{238}{100}$
6 0·7
7 1·5
8 0·07
9 2·7
10 0·16

Number Splitting for Division

1

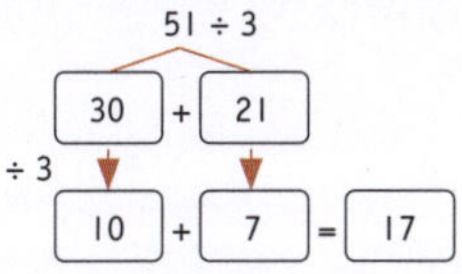

2

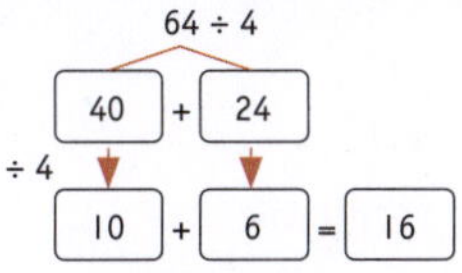

3

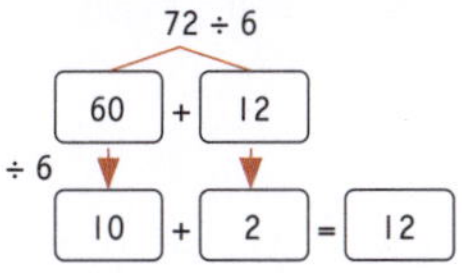

4

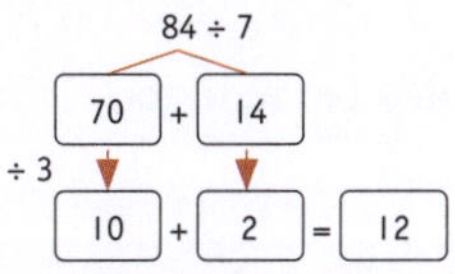

5

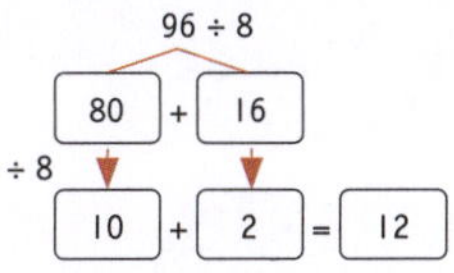

6

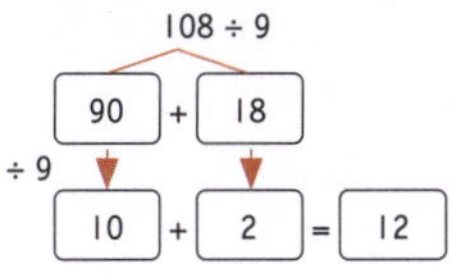

ANSWERS

UNIT 11 STATISTICS & PROBABILITY

Just by Chance

1 Likely
2 There are the same number of red as black cards in the pack. There is always an even chance of either colour being drawn.

UNIT 11 MEASUREMENT & SPACE

Decimal Measurement

1 1·35 kg	8 1700 mL	15 9.8 m
2 2·22 L	9 1060 g	16 3420 g
3 1·2 km	10 3400 mL	17 104 cm
4 1·3 m	11 2.65 kg	18 52 600 mL
5 3·5 m	12 0.763 L	19 40 g
6 1650 g	13 1.2 km	20 21 700 mL
7 340 cm	14 21.4 m	

UNIT 11 NAPLAN*-STYLE QUESTIONS

1 0·9　　2 3　　3 500 mL

UNIT 11 PROBLEM OF THE WEEK

Making a Flow Chart

King of Diamonds

UNIT 12 NUMBER & ALGEBRA

Multiplying by 10 and 100

1 11	3 34·6	5 100	7 134
2 0·1	4 823	6 11	8 706

Quick Check

1 8	3 13	5 25·7	7 75	9 103
2 7·5	4 10·3	6 80	8 130	10 257

Rounding Money Amounts

	Item price	Rounded price
1	$1·62	$1·60
	$2·35	$2·40
	$3·99	$4·00
	$4·15	$4·00
Total	$12·11	$12·20

	Item price	Rounded price
3	$3·55	$3·60
	$4·64	$4·60
	$2·23	$2·20
	$5·12	$5·10
Total	$15·54	$15·50

	Item price	Rounded price
2	$1·75	$1·80
	$2·95	$3·00
	$1·15	$1·20
	$2·03	$2·00
Total	$7·88	$8·00

	Item price	Rounded price
4	$3·45	$3·50
	$2·19	$2·20
	$3·94	$3·90
	$1·22	$1·20
Total	$10·80	$10·80

UNIT 12 STATISTICS & PROBABILITY

A Sales Comparison

1 Craft　　2 Cooking　　3 Cooking　　4 5000

UNIT 12 MEASUREMENT & SPACE

Making Characters with Shapes

Open ended

UNIT 12 NAPLAN*-STYLE QUESTIONS

1 1·75　　2 $9　　3 6 triangles

UNIT 12 PROBLEM OF THE WEEK

Screen Time

The second parts of these questions are open ended – suggestions only given below.

1 Between 6:30 and 7:00. Graph 1 because I could see the answer on the top line.
2 Friday. Graph 2, because it shows how many hours each day.
3 Wednesday. Graph 2, because the scale shows the answer.
4 Tuesday. Graph 2, because the scale shows the highest point.

UNIT 13 NUMBER & ALGEBRA

Decimal Notation and Money

1 $2·35	3 $3·75	5 $1·53	7 $14
2 $5·05	4 $12·10	6 $2·05	8 $23·45

Adding Cents

9 $1·05	11 $1·15	13 $5·10	15 $11·10
10 $1·70	12 $2·20	14 $9·60	16 $22

Quick Check

1 $1·55	3 $3·05	5 $2·45	7 $1·10	9 $3·00
2 $3·50	4 $1·25	6 $1·03	8 $2·10	10 $2·05

Multiplication and Division Fact Families

1 Use 7 × 5 = 35 for 35 ÷ 7 = 5 and 35 ÷ 5 = 7
2 Use 8 × 4 = 32 for 32 ÷ 8 = 4 and 32 ÷ 4 = 8
3 Use 6 × 9 = 54 for 54 ÷ 6 = 9 and 54 ÷ 9 = 6
4 Use 9 × 7 = 63 for 63 ÷ 9 = 7 and 63 ÷ 7 = 9
5 Use 4 × 9 = 36 for 36 ÷ 4 = 9 and 36 ÷ 9 = 4
6 Use 7 × 6 = 42 for 42 ÷ 7 = 6 and 42 ÷ 6 = 7
7 Use 8 × 6 = 48 for 48 ÷ 8 = 6 and 48 ÷ 6 = 8
8 Use 7 × 3 = 21 for 21 ÷ 7 = 3 and 21 ÷ 3 = 7
9 Use 3 × 12 = 36 for 36 ÷ 3 = 12 and 36 ÷ 12 = 3
10 Use 8 × 9 = 72 for 72 ÷ 8 = 9 and 72 ÷ 9 = 8

UNIT 13 STATISTICS & PROBABILITY

How well do you know your 9 times table?

Open ended

UNIT 13 MEASUREMENT & SPACE

Decimals for Measuring

1 1·35 m
2 1·05 m
3 3·3 m (The zero in 330 cm is not needed in the metre form)
4 2·5 m
5 47 m
6 0·12 m
7 2·35 m
8 1·64 m
9 4·05 m
10 7·6 m
11 12·4 m
12 0·82 m

UNIT 13 NAPLAN*-STYLE QUESTIONS

1 8　　2 $4·35　　3 Unlikely

UNIT 13 PROBLEM OF THE WEEK

A Minute Game

1 Closest is 25 − 3 = 22
2 Closest is (3 + 4) × 5 = 35
3 Open ended

UNIT 14 NUMBER & ALGEBRA

Odd and Even Numbers

1 E, O, E, O, E, O, E
2 O, O, O, O, O, O, O
3 O, E, O, E, O, E, O
4 E, O, E, O, E, O, E
5 E, O, E, O, E, O, E
6 E, E, E, E, E, E, E
7 E, E, E, E, E, E, E
8 O, O, O, O, O, O, O
9 E, E, E, E, E, E, E
10 O, O, O, O, O, O, O

Quick Check

1 Odd	3 Odd	5 Even	7 Odd	9 Even
2 Even	4 Odd	6 Even	8 Odd	10 Even

Equivalent Additions

1 5	3 16	5 4	7 8	9 10
2 11	4 6	6 15	8 22	10 3

TARGETING MATHS HOMEWORK: YEAR 4 © PASCAL PRESS ISBN: 9781925726572

UNIT 14 STATISTICS & PROBABILITY

What Chance of Going to the Movies?

1 Certain
2 Very likely
3 Likely
4 Unlikely
5 Yes, because things (or circumstances) change.

UNIT 14 MEASUREMENT & SPACE

Where is X?

X is in D3, under the palm trees:

UNIT 14 NAPLAN*-STYLE QUESTIONS

1 odd 2 10 3 Likely

UNIT 14 PROBLEM OF THE WEEK

In Rows and Circles

Following Kai and Jo's hint, start by adding the numbers 2 to 10 which gives 54. Each number will appear in two lines or in a line and the circle. So each number will be used twice in the totals and there are four totals (3 lines and 1 circle) to make. 54 + 54 = 108, 108 ÷ 4 = 27.

If the circle numbers add to 27, that leaves 54 − 27 = 27 for the three numbers in the middle which must add to 27 and can only be 8, 9 and 10.

Now try to complete the line that has 8 and 9. 8 + 9 = 17, and so the two circle numbers on that line must add to 10. There are two possibilities and both lead to successful solutions:

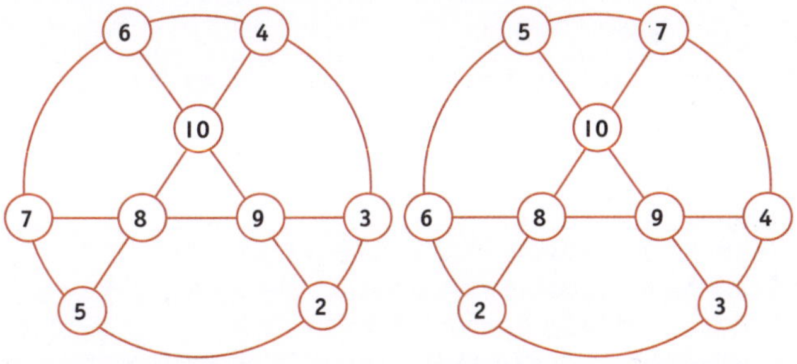

Congratulations to every student who successfully completes this Problem of the Week!

UNIT 15 NUMBER & ALGEBRA

Landmark Numbers

1 50 cm or 0·5 m
2 25 cm or 0·25 m
3 75 cm or 0·75 m
4 150 cm or 1·5 m
5 125 cm or 1·25 m
6 50 cm or 0·5 m
7 150 cm or 1·5 m
8 150 cm or 1·5 m
9 175 cm or 1·75 m
10 75 cm or 0·75 m

Quick Check

1 $\frac{3}{4}$ of a metre
2 $1\frac{1}{4}$ m
3 50 cm
4 0·25 metres
5 0·5 m of elastic or $\frac{1}{2}$ m of elastic

Multiplying by 9

1 6 × 9 = 6 × 10 − 6 = 54
2 8 × 9 = 8 × 10 − 8 = 72
3 13 × 9 = 13 × 10 − 13 = 117
4 20 × 9 = 20 × 10 − 20 = 180
5 16 × 9 = 16 × 10 − 16 = 144
6 35 × 9 = 35 × 10 − 35 = 315
7 19 × 9 = 19 × 10 − 19 = 171
8 26 × 9 = 26 × 10 − 26 = 234
9 23 × 9 = 23 × 10 − 23 = 207
10 89 × 9 = 89 × 10 − 89 = 801

UNIT 15 STATISTICS & PROBABILITY

Kitty's Growth Chart

1 100 g
2 Thursday
3 Friday
4 Saturday
5 60 g
6 10 g

UNIT 15 MEASUREMENT & SPACE

Measuring Area

1 10 squares
2 16 squares
3 22 squares
4 20 squares
5 16 squares
6 10 squares
7 9 squares

UNIT 15 NAPLAN*-STYLE QUESTIONS

1 0·75 2 12 × 10 − 12 3 7 squares

UNIT 15 PROBLEM OF THE WEEK

The Farmer's Hurdles

Here are some possible ways of enclosing 6 squares with 12 hurdles.

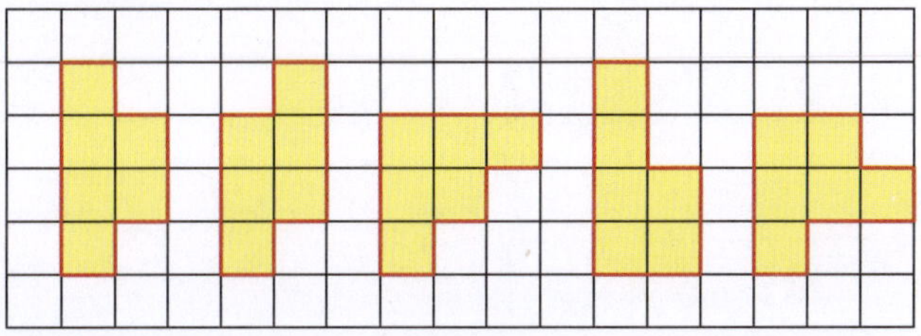

Note: All the above are made by starting with a simple 4-square and adding two more squares to the sides as shown below:

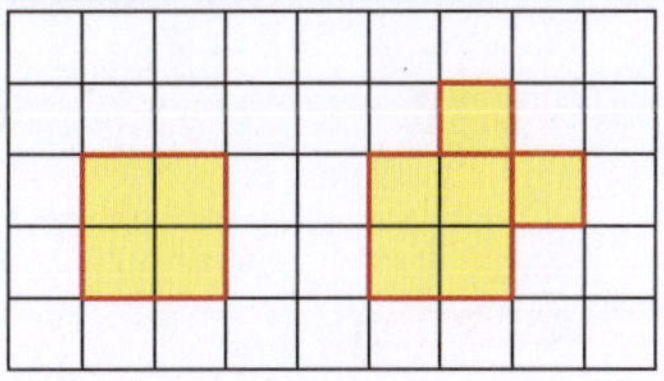

UNIT 16 NUMBER & ALGEBRA

Counting in Decimals

1 2·9, 2·7. Decreasing by 0·2
3·3 − 0·2 = = =

2 3·7, 4. Increasing by 0·3
2·8 + 0·3 = = = =

3 2·7, 2·2. Decreasing by 0·5
4·2 − 0·5 = = = =

4 5·75, 6, 6·25. Increasing by 0·25
5 + 0·25 = = = = =

5 9·4, 10·5, 11·6. Increasing by 1·1
6·1 + 1·1 = = = = =

Quick Check

1 5·9
2 2·9
3 0·1
4 2·7
5 4·23
6 0·85
7 1·75
8 1·22
9 1·00
10 0·85

Word Problems

1 10 × 65 c = $6·50
2 3 × 25 cm = 75 cm
3 9 × $25 = $225
4 100 × $5·25 = $525
5 $3 \div \frac{1}{3} = 9$

UNIT 16 STATISTICS & PROBABILITY

Weather Chart

	Impossible	Unlikely	Likely	Very likely	Certain
Mon.	✔				
Tue.		✔			
Wed.				✔	
Thu.		✔			
Fri.					✔
Sat.			✔		
Sun.		✔			

UNIT 16 MEASUREMENT & SPACE

Symmetry

1
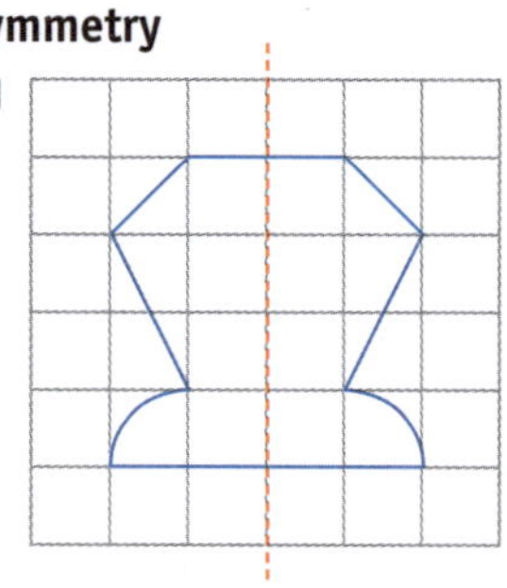

3
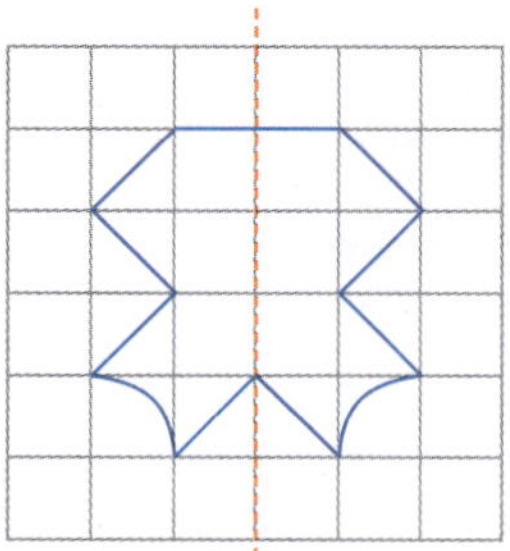

2
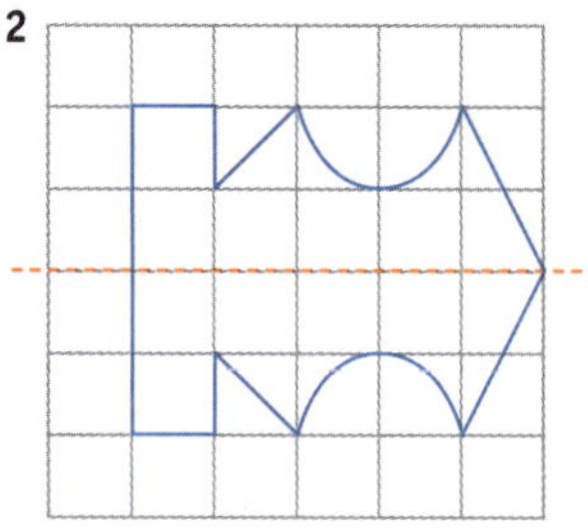

4
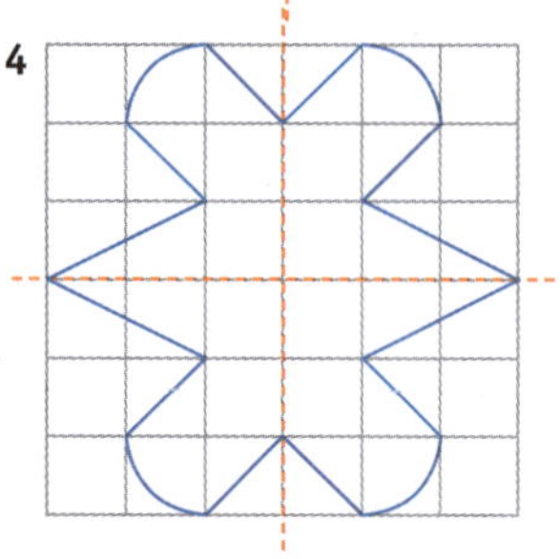

UNIT 16 NAPLAN*-STYLE QUESTIONS

1 $13
2 4·1
3
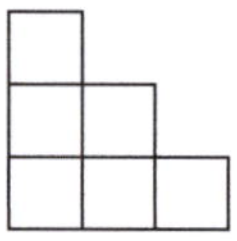

UNIT 16 PROBLEM OF THE WEEK

Symmetry with 7 Tiles

Here are some possible answers that use the hint.
Also included is one that wins a bonus mark.

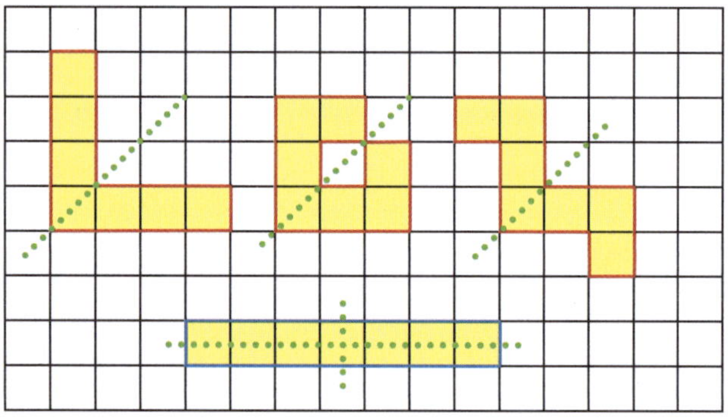

TERM 2 REVIEW

Review: Unit 9 and Unit 10

1 0·7
2 1
3 2·3
4 0·5
5 1·2
6 2·9
7 $2\frac{1}{4}$, $2\frac{2}{4}$, $2\frac{3}{4}$, 3
8 $\frac{5}{3}$, $\frac{6}{3}$, $\frac{7}{3}$, $\frac{8}{3}$, $\frac{9}{3}$
9 15 people like Brussels sprouts.

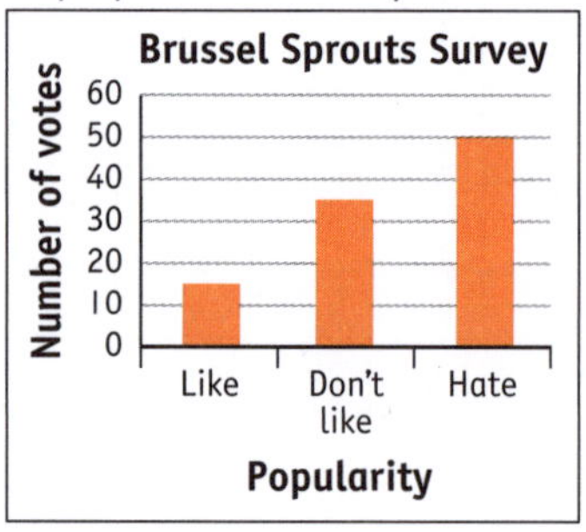

10 $\frac{6}{10}$
11 $\frac{77}{100}$
12 $\frac{4}{10}$ or $\frac{40}{100}$
13 $\frac{9}{10}$ or $\frac{90}{100}$

14

×	30	+ 6	
5	150	30	= 180

15

×	200	+ 30	+ 4	
6	1200	180	24	= 1404

16 14 °C

Review: Unit 11 and Unit 12

1 $1\frac{36}{100}$
2 $1\frac{5}{100}$
3 $2\frac{73}{100}$
4 1·8
5 11·2
6 2·08

7
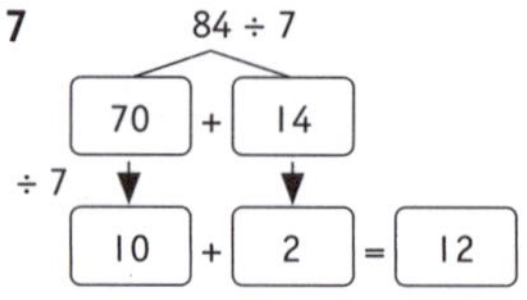

8
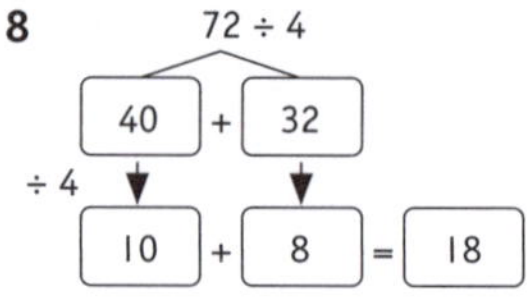

9 Unlikely
10 1·123 kg
11 1500 mL
12 13·4
13 20·5
14 134
15 205

16

	Item price	Rounded price
	$1·54	$1·50
	$2·75	$2·80
	$3·69	$3·70
	$2·28	$2·30
Total	$10·26	$10·30

17

	Item price	Rounded price
	$3·25	$3·00
	$4·59	$5·00
	$6·22	$6·00
	$7·53	$8·00
Total	$21·59	$22·00

Review: Unit 13 and Unit 14

1 $1·65
2 $3·05
3 $2·20
4 Use 5 × 3 = 15 for 15 ÷ 5 = 3 and 15 ÷ 3 = 5
5 Use 6 × 8 = 48 for 48 ÷ 6 = 8 and 48 ÷ 8 = 6
6 Use 9 × 7 = 63 for 63 ÷ 9 = 7 and 63 ÷ 7 = 9
7 2·64 m
8 3·08 m
9 Odd
10 Even
11 Even
12 Odd
13 Odd
14 Even
15 17 + 3 = 7 + 13
16 29 + 6 = 30 + 5
17 5 + 17 = 17 + 5
18 6 + 27 = 16 + 17
19 Impossible
20 Certain
21 Likely
22 Certain

Review: Unit 15 and Unit 16

1 0·75 m
2 1·25 m
3 7 × 9 = 7 × 10 − 7 = 63
4 14 × 9 = 14 × 10 − 14 = 126
5 Saturday
6 Wednesday
7 8 °C
8 11 squares
9 9 squares
10 3·9, 4·1, 4·3
11 3·2, 2·9, 2·6
12 6 × 4 = 24
13 $\frac{1}{4} + \frac{1}{2} = \frac{3}{4}$

14
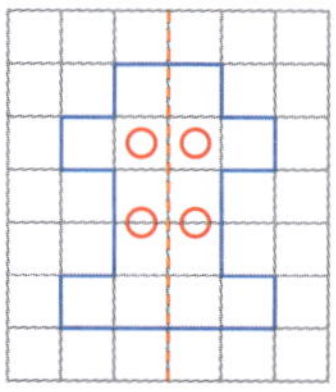

15
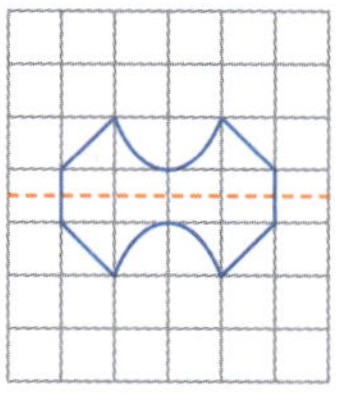

TERM 3

UNIT 17 NUMBER & ALGEBRA

Estimating Money Amounts

		Number	Value
2	$2	1	$2
3	$1	2	$2
4	50c	6	$3
5	20c	8	$1·60

		Number	Value
6	10c	4	40c
7	5c	10	50c
8	Total	31	$9·50

ANSWERS

Quick Check

1 $12	3 $1·60	5 $1·05	7 $5·50	9 $3·20
2 $2·50	4 $1·50	6 $5·20	8 $2	10 $7·80

Fractions (proper, improper and mixed)

1 $\frac{3}{4}, \frac{7}{8}, 1\frac{1}{8}, \frac{5}{4}, \frac{13}{8}$

2 $\frac{1}{6}, \frac{1}{3}, \frac{3}{6}, \frac{7}{6}, \frac{4}{3}$

3 $\frac{1}{2}, \frac{3}{5}, \frac{11}{10}, 1\frac{1}{5}, \frac{8}{5}$

4 $\frac{1}{10}, \frac{1}{8}, \frac{1}{6}, \frac{1}{5}, \frac{1}{4}$

5 $\frac{3}{5}, \frac{3}{4}, 1\frac{1}{5}, 1\frac{1}{4}, 1\frac{1}{2}$

UNIT 17 STATISTICS & PROBABILITY

Votes for an Excursion

1 No
2 Beach volleyball
3 Museum
4 7
5 No. Not many girls liked it.

UNIT 17 MEASUREMENT & SPACE

Time Durations

The diagrams are open ended, but the times are not.

1 Total Time = 2 h 20 min
2 Total Time = 2 h 55 min
3 Cooked at 12:25
4 Total Time = 57 min
5 Total Time = 6 h 26 min

UNIT 17 NAPLAN*-STYLE QUESTIONS

1 $2·75　2 $1\frac{1}{3}$　3 85 min

UNIT 17 PROBLEM OF THE WEEK

Checking Their Savings

The totals for Claire and Jake need to be worked out:

Claire had:	
18 × 5c	$0·90
12 × 10c	$1·20
22 × 20c	$4·40
3 × 50c	$1·50
Total	$8·00

Jake had:	
23 × 5c	$1·15
15 × 10c	$1·50
18 × 20c	$3·60
6 × 50c	$3
Total	$9·25

1 Jo had $3·75.
2 Jo had 7 × 50c, 1 × 20c, 0 × 10c and 1 × 5c
3 $4·25
4 $5·50
5 $7·00
5 Claire gives Jo $1 so that Claire has exactly $7. The amount must be made from the coins that Claire has. Hence 2 × 50c.
6 Jake gives Jo $2·25 so that Jake has exactly $7. The amount must be made from the coins that Jake has. Hence 4 × 50c, 1 × 20c and 1 × 5c.

UNIT 18 NUMBER & ALGEBRA

Extending Multiplication Facts

1 400	3 490	5 320	7 630	9 360
2 420	4 810	6 560	8 560	10 440

Quick Check

1 840	3 980	5 1320	7 210	9 250
2 760	4 1020	6 240	8 320	10 630

A Strategy for Multiplying

1 56　2 45　3 112　4 63　5 168

UNIT 18 STATISTICS & PROBABILITY

Top Holiday Destination

1

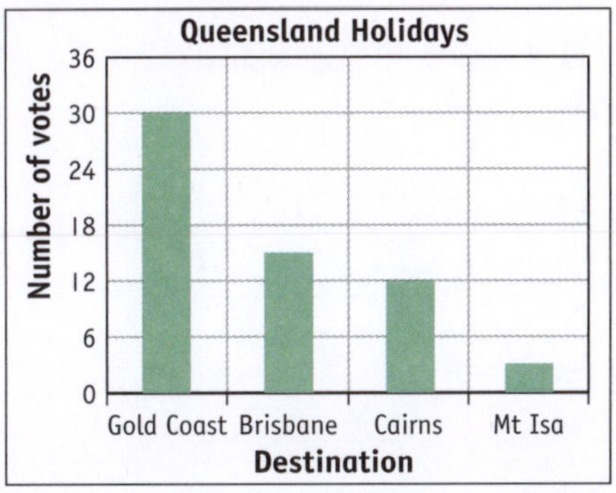

2 15　3 30　4 12　5 60

UNIT 18 MEASUREMENT & SPACE

Area Calculations

1 6 cm²　2 20 cm²　3 18 cm²　4 12 cm²　5 18 cm²

UNIT 18 NAPLAN*-STYLE QUESTIONS

1 20
2 20
3 Ask every student in the class.

UNIT 18 PROBLEM OF THE WEEK

Flood Damage

1 56 carpet tiles　2 $392　3 $108

UNIT 19 NUMBER & ALGEBRA

Extending Division Facts

1 70	3 70	5 80	7 120
2 70	4 60	6 80	8 90

Quick Check

1 78	3 94	5 104	7 50	9 70
2 56	4 32	6 90	8 40	10 20

Finding Unknown Values in Additions

1 8	4 23	7 47	10 54
2 4	5 19	8 64	11 85
3 21	6 62	9 200	12 810

UNIT 19 STATISTICS & PROBABILITY

Drop a Coin

Open ended

UNIT 19 MEASUREMENT & SPACE

Lines of Symmetry

1 2 lines of symmetry
2 0 lines of symmetry
3 3 lines of symmetry
4 Open ended
5 Open ended

UNIT 19 NAPLAN*-STYLE QUESTIONS

1 116　2 Triangle　3 Second shape

UNIT 19 PROBLEM OF THE WEEK

Make a Logo Design

Open ended

UNIT 20 NUMBER & ALGEBRA

Rounding Decimals

1 4	6 4	11 7·6	16 10	21 41
2 4	7 13	12 9	17 6	22 24
3 4	8 15	13 16·3	18 12	23 20
4 5	9 20	14 12·6	19 9	24 50
5 4	10 6·8	15 19·7	20 8	

Quick Check

1 5	3 8	5 3	7 6·4	9 1·7
2 5	4 6	6 4·9	8 3·5	10 2

Finding Unknown Values in Subtractions

1 17 2 15 3 39 4 46 5 19 6 56 7 81 8 39 9 16

UNIT 20 STATISTICS & PROBABILITY

The Class Car Colour Survey

1 Claire's Graph

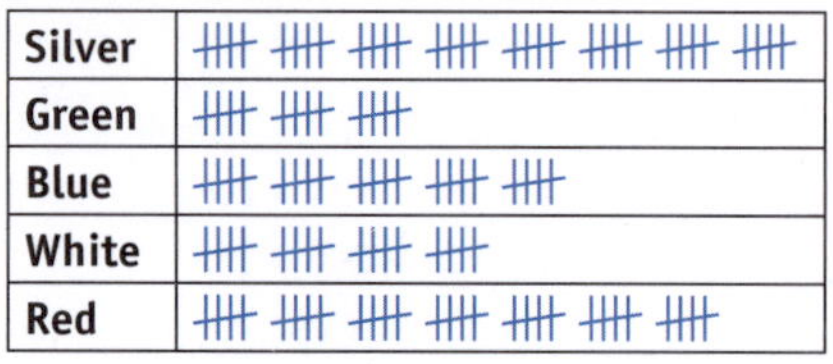

Chen's Graph

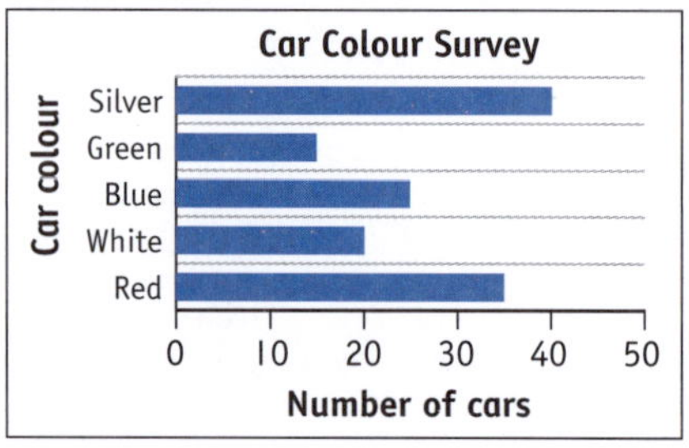

Zoe's Graph

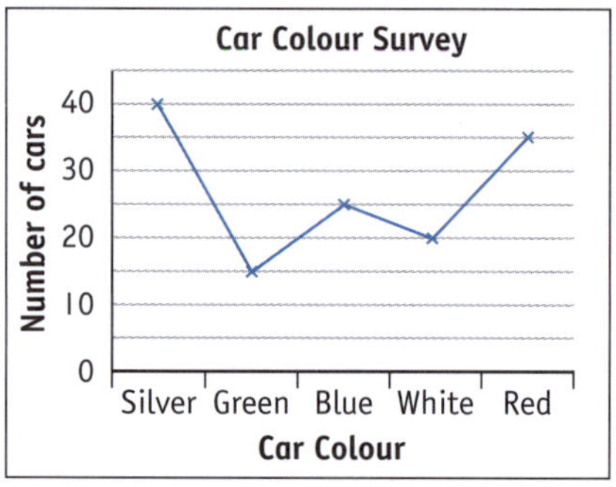

2 135, using tallies
3 5, using the line graph
4 Red, blue, white, using the line graph
5 15, using the bar graph

UNIT 20 MEASUREMENT & SPACE

About Time

1 Twenty-seven past two, 2:27, right angle
2 Twenty-two minutes to two, 1:38, straight angle
3 Half past four, 4:30, acute angle
4 Twenty to one, 12:40, reflex angle
5 Quarter past seven, 7:15, obtuse angle
6 Eleven past eleven, 11:11, right angle

UNIT 20 NAPLAN*-STYLE QUESTIONS

1 0·75 2 A 3 37

UNIT 20 PROBLEM OF THE WEEK

Make to 10

3·29 + 6·71 = 10 4·09 + 5·91 = 10
4·55 + 5·45 = 10 6·75 = 3·25 = 10
6·02 + 3·98 = 10 7·25 + 2·75 = 10
4·51 + 5·49 = 10 6·95 + 3·05 = 10

The number on the covered card is 3·05.

UNIT 21 NUMBER & ALGEBRA

Equivalent Fractions

1 $\frac{2}{2} = 1$
2 $\frac{6}{6} = \frac{2}{2} = 1$
3 $\frac{3}{6} = \frac{1}{2}$, $\frac{2}{6} = \frac{1}{3}$
4 $\frac{3}{3} = 1$
5 $\frac{3}{3} = \frac{9}{9} = 1$
6 $\frac{3}{9} = \frac{1}{3}$, $\frac{6}{9} = \frac{2}{3}$
7 $\frac{5}{5} = 1$
8 $\frac{5}{5} = \frac{10}{10} = 1$
9 $\frac{2}{10} = \frac{1}{5}$, $\frac{5}{10} = \frac{1}{2}$

Quick Check

1 $\frac{6}{8}$ 2 $\frac{1}{4}$ 3 $\frac{8}{10}$ 4 $\frac{1}{3}$ 5 $\frac{2}{3}$ 6 $\frac{1}{2}$ 7 $\frac{1}{2}$ 8 $\frac{2}{5}$

Finding Unknown Values in Money Problems

1

$4·50	
$1·50	$3

2

$12·70		
$2·20	$4·50	$6

3

$30		
$17	$6	$7

4

$86		
$57	$9	$20

UNIT 21 STATISTICS & PROBABILITY

Just by Chance

1 10
2 Yellow
3 All are likely.
4 There are about the same number of each colour.

UNIT 21 MEASUREMENT & SPACE

The Unfinished Map

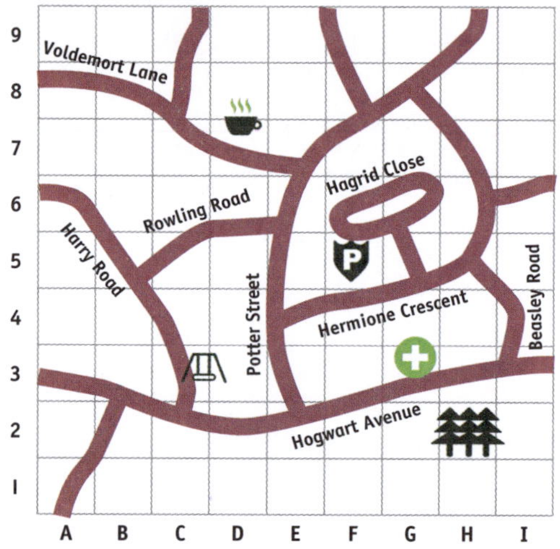

UNIT 21 NAPLAN*-STYLE QUESTIONS

1 $5·25 2 $\frac{6}{8}$ 3 East

UNIT 21 PROBLEM OF THE WEEK

The Darts Tournament

Player 1 scores were 48, 32 and 20
Player 2 scores were 60, 21 and a miss.
Player 3 scores were 50, 36 and 17.
Extra Challenge: Open ended

UNIT 22 NUMBER & ALGEBRA

Addition by the Front-end Method

1
```
   26
 + 39
 ----
   50
 + 15
 ----
   65
```

2
```
   84
 + 23
 ----
  100
 +  7
 ----
  107
```

3
```
   37
 + 56
 ----
   80
 + 13
 ----
   93
```

4
```
   75
 + 36
 ----
  100
 + 11
 ----
  111
```

5
```
  136
+ 223
-----
  300
   50
+   9
-----
  359
```

6
```
  275
+ 332
-----
  500
  100
+   7
-----
  607
```

7
```
  436
+ 175
-----
  500
  100
+  11
-----
  611
```

8
```
  647
+ 485
-----
 1000
  120
+  12
-----
 1132
```

ANSWERS

TARGETING MATHS HOMEWORK: YEAR 4 © PASCAL PRESS ISBN: 9781925726572

Quick Check

1 59 3 107 5 99 7 51 9 147
2 87 4 128 6 61 8 44 10 171

Rounding Money

1	3 loaves at $3·85 each	3 @ $4	$12	$11·55
2	6 lettuces at $3·45 each	6 @ $3·50	$21	$20·70
3	2 chickens at $8·20 each	2 @ $8	$16	$16·40
4	4 avocados at $1·45 each	4 @ $1·50	$6	$5·80
5	3 cucumbers at $1·95 each	3 @ $2	$6	$5·85
6	Altogether about:		$61	
7	Now check the actual amounts with your calculator.			$60·30
8	Was the estimate close? Yes			
9	Was the estimate higher or lower? Higher			
10	Why do you think that was? Open ended			

UNIT 22 STATISTICS & PROBABILITY

The Dalmatian Puppies

1 Unlikely
2 Very likely
3 Unlikely
4 Impossible
5 Impossible

UNIT 22 MEASUREMENT & SPACE

Finding Perimeters

1 20 + 11·5 = 31·5, 31·5 × 2 = 63 metres
2 6·5 × 4 = 26 metres
3 12 × 4 = 48 cm
4 75 + 60 = 135 cm, 135 × 2 = 2·7 metres
5 135 + 125 = 260, 260 × 2 = 5·2 metres

UNIT 22 NAPLAN*-STYLE QUESTIONS

1 $601
2 16 m
3 It could be either.

UNIT 22 PROBLEM OF THE WEEK

A Rainy Day Game

First Round
1 863
2 128, 242 and 276
3 853
4 Zoe

Second Round
5 128, 225 and 276
6 225, 253 and 334
7 595
8 Zoe with 1492
9–11 are open ended.

UNIT 23 NUMBER & ALGEBRA

Pattern Rules

1 36, 72, 75, 150, 153
2 23, 20, 40, 37, 74
3 50, 25, 30, 15, 20
4 20, 100, 95, 475, 470
5 60, 300, 290, 1450, 1440
6 Multiply by 2, add 2.
28, 56, 58, 116, 118
7 Multiply by 4, add 4.
9, 36, 40, 160, 164
8 Add 10, multiply by 2.
14, 24, 48, 58, 116
9 Add 9, multiply by 4.
4, 13, 52, 61, 244
10 Multiply by 6, subtract 10.
5, 30, 20, 120, 110
11 Multiply by 2, add 5.
12 Subtract 3, add 5.
13 Divide by 2, subtract 1.
14 Add 1, multiply by 3.

Quick Check

1 36, 39, 39
2 74, 71, 142
3 21, 63, 61
4 32, 16, 22
5 27, 9, 0
6 subtract 11
7 divide by 3
8 multiply by 2
9 divide by 10
10 multiply by 9

Adding with Decimals

1 7, $\frac{11}{10}$ or 1·1, 8·1
2 10, $\frac{12}{10}$ or 1·2, 11·2
3 7, $\frac{15}{10}$ or 1·5, 8·5
4 9 + 1·6 = 10·6
5 27 + 0·9 = 27·9
6 30 + 1·3 = 31·3

UNIT 23 STATISTICS & PROBABILITY

A Shoe Shuffle

1

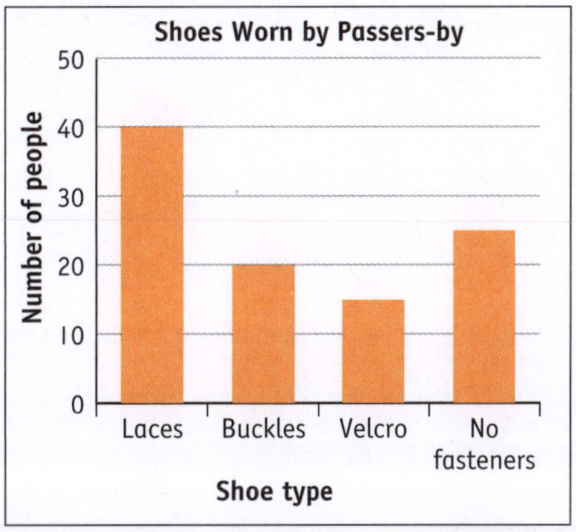

2 40 3 20 4 15 5 25

UNIT 23 MEASUREMENT & SPACE

Finding Areas

1 3 × 6 = 18
2 4 × 6 = 24
3 5 × 4 = 20
4 7 × 6 = 42
5 3 × 8 = 24
6 7 × 7 = 49

UNIT 23 NAPLAN*-STYLE QUESTIONS

1 3·4 °C 2 3·4 3 21

UNIT 23 PROBLEM OF THE WEEK

Making Rectangles

Rectangles can be 1 × 24, 2 ×12, 3 × 8, or 4 × 6.
1 Open ended
2 50 – 20 = 30
3 1 × 12
4 Open ended
5 Should show that 1 × 12 has the longest perimeter.

UNIT 24 NUMBER & ALGEBRA

Fractions and Division

The diagrams should represent the equation shown.

1 3 × 4 = 12
2 4 × 3 = 12
3 12 ÷ 3 = 4
4 12 ÷ 4 = 3
5 $\frac{1}{3}$ of 12 = 4
6 $\frac{1}{4}$ of 12 = 3
7 3 × 5 = 15
8 5 × 3 = 15
9 15 ÷ 3 = 5
10 15 ÷ 5 = 3
11 $\frac{1}{3}$ of 15 = 5
12 $\frac{1}{5}$ of 15 = 3

Quick Check

1 18 ÷ 3, 18 ÷ 6
2 20 ÷ 4, 20 ÷ 5
3 21 ÷ 3, 21 ÷ 7
4 48 ÷ 8, 48 ÷ 6
5 45 ÷ 9, 45 ÷ 5
6 $\frac{1}{6}$ of 18 = 3 or $\frac{1}{3}$ of 18 = 6
7 $\frac{1}{5}$ of 20 = 4 or $\frac{1}{4}$ of 20 = 5
8 $\frac{1}{3}$ of 21 = 7 or $\frac{1}{7}$ of 21 = 3
9 $\frac{1}{8}$ of 48 = 6 or $\frac{1}{6}$ of 48 = 8
10 $\frac{1}{5}$ of 45 = 9 or $\frac{1}{9}$ of 45 = 5

Finding the Difference

1 28
2 18
3 28
4 29
5 102
6 23

UNIT 24 STATISTICS & PROBABILITY

Training for the Triathlon

1 15 2 46 3 0 4 4 5 9

UNIT 24 MEASUREMENT & SPACE

Calculating Mass

1 10 apples with mass 2 kg
2 4 apples with mass 800 g
3 15 apples with mass 3 kg
4 2 apples with mass 400 g

UNIT 24 NAPLAN*-STYLE QUESTIONS

1 10 2 Pumpkin 3 7

ANSWERS

UNIT 24 PROBLEM OF THE WEEK

Difference Squares

The most likely outcome of choosing numbers 0–9 is to make all the numbers on the inner square equal to 0. However, 2 and 4 can also appear. In this example, the even numbers are used in the start positions:

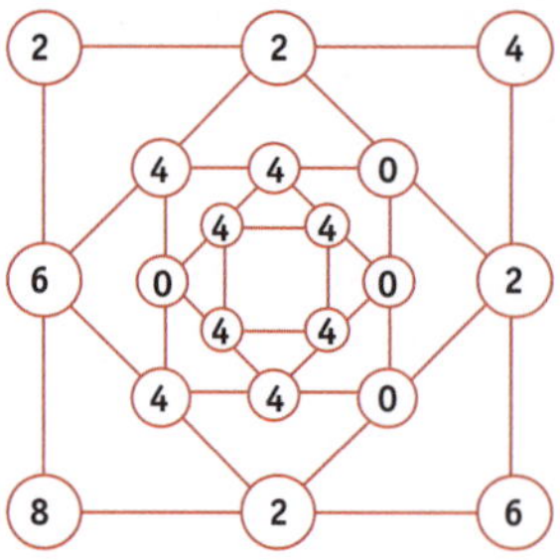

If larger numbers (0–20) are used in the start positions, two more even numbers can appear in the final square.
The pattern is that 0 is the number that is most likely to appear, but even numbers 2, 4 , 6 and 8 can also appear. The numbers on the finish square are always even numbers.

TERM 3 REVIEW

Review: Unit 17 and Unit 18

1 $2·50
2 70 cents
3 $1·45
4 $2·40
5 $10
6 $\frac{3}{8}$, $\frac{3}{4}$, $1\frac{1}{8}$, $1\frac{1}{4}$, $\frac{7}{4}$
7 $\frac{4}{4}$, $\frac{9}{8}$, $\frac{5}{4}$, $1\frac{1}{2}$, $\frac{4}{2}$
8 1 h 55 min
9 1 h 45 min
10 280
11 250
12 720
13 63
14 72
15 64
16 9
17 54
18 Area = 12 squares
19 Area = 14 squares

Review: Unit 19 and Unit 20

1 72
2 80
3 49
4 50
5 90
6 80
7 19
8 253
9 29
10 Likely
11

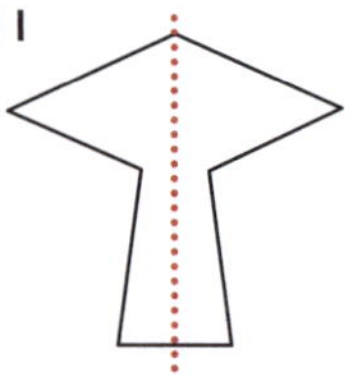

12

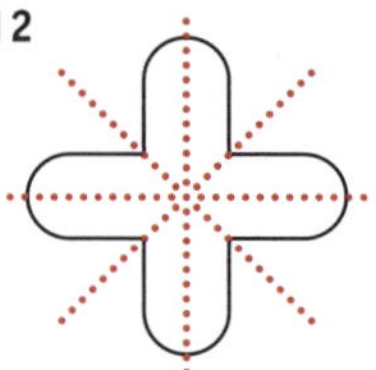

13

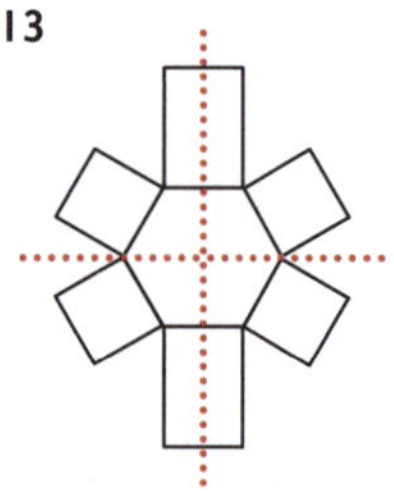

14 7
15 4
16 6
17 3·8
18 2·8
19 4
20 17
21 54
22 22
23 straight angle, 6 o'clock, 6:00
24 acute angle, quarter to 7, 6:45
25 reflex angle, 20 to 1, 12:40

Review: Unit 21 and Unit 22

1 $\frac{1}{2}$
2 $\frac{2}{3}$
3 $\frac{2}{5}$
4 $\frac{6}{8}$
5 $17 + $28 = $45

45	
17	28

6 $57 + $15 = $72

72	
57	15

7 $56 – $37 = $19

56	
19	37

8 Unlikely. It is just possible that there is a black counter in the bag.

9
```
   28
 + 69
   80
 + 17
   97
```

10
```
   79
 + 36
  100
 + 15
  115
```

11
```
  126
+ 364
  400
   80
 + 10
  490
```

12
```
  653
+ 374
  900
  120
 +  9
 1027
```

13 $10, $9·75
14 $9, $8·55
15 Likely
16 8 cm × 8 cm = 64 cm²
17 297 mm + 210 mm
= 507 mm
507 mm × 2 = 1014 mm
= 1.014 m

Review: Unit 23 and Unit 24

1 6, 12, 15, 30, 33
2 8, 6, 18, 16, 48
3 68, 34, 40, 20, 26
4 6, 24, 21, 84, 81
5 5·8
6 8·1
7 10·5
8 2 × 6 = 12 squares
9 3 × 5 = 15 squares
10 3 × 5 = 15
11 15 ÷ 3 = 5
12 $\frac{1}{5}$ of 15 = 3
13 28 ÷ 7 = 4 and 28 ÷ 4 = 7
14 54 ÷ 9 = 6 and 54 ÷ 6 = 9
15 8
16 34
17 250 g
18 2 kg
19 6
20 3

TERM 4

UNIT 25 NUMBER & ALGEBRA

Rounding and Estimating for Addition

1 20 + 50 = 70, 30 + 60 = 90.
The range is between 70 and 90.
2 40 + 60 = 100, 50 + 70 = 120.
The range is between 100 and 120.
3 70 + 50 = 120, 80 + 60 = 140.
The range is between 120 and 140.
4 90 + 60 = 150, 100 + 70 = 170.
The range is between 150 and 170.
5 70 + 60 = 130, 80 + 70 = 150.
The range is between 130 and 150.
6 20 + 30 + 40 = 90, 30 + 40 + 50 = 120.
The range is between 90 and 120.
7 90 + 30 + 40 = 160, 100 + 40 + 50 = 190.
The range is between 160 and 190.
8 30 + 20 + 100 = 150, 40 + 30 + 110 = 180.
The range is between 150 and 180.
9 40 + 60 + 70 = 170, 50 + 70 + 80 = 200.
The range is between 170 and 200.
10 70 + 120 + 50 = 240, 80 + 130 + 60 = 270.
The range is between 240 and 270.

Quick Check

1 90 to 110
2 150 to 170
3 120 to 140
4 170 to 190
5 160 to 180
6 130 to 150
7 50 to 70
8 140 to 160
9 210 to 230
10 310 to 330

Rounding and Estimating Money Amounts

1 $4 + $3 + $4 = $11. A safe estimate is $11.
2 $5 + $3 + $7 = $15. A safe estimate is $15.
3 $11 + $7 + $4 = $22. A safe estimate is $22.
4 $16 + $4 + $12 = $32. A safe estimate is $32.
5 $9 + $4 + $18 = $31. A safe estimate is $31.
6 $27 + $7 + $6 = $40. A safe estimate is $40.

UNIT 25 STATISTICS & PROBABILITY

Under the Cups

Open ended.

TARGETING MATHS HOMEWORK: YEAR 4 © PASCAL PRESS ISBN: 9781925726572

UNIT 25 MEASUREMENT & SPACE

All about Cylinders

Open ended.

UNIT 25 NAPLAN*-STYLE QUESTIONS

1 $1478
2 cylinder and rectangular prism
3 6 sixes

UNIT 25 PROBLEM OF THE WEEK

Different Differences

Examples are {1, 2, 4}, {1, 2, 5}, {1, 2, 6} or {1, 3, 6}, {1, 3, 7}, {1, 3, 8}.
Each of these sets can be extended with another digit to solve the second part, but to make the largest number as small as possible, make sure that the fourth number is twice the third number in the sets of three numbers. Examples are {1, 2, 4, 8}, {1, 2, 5, 10} or {1, 3, 6, 12}, {1, 3, 7, 14}.

UNIT 26 NUMBER & ALGEBRA

Estimating Money Amounts

1 8 at $24·50, 8 at $25 = $200, 10 at $24·50 = $245
Actual Cost $196
2 9 at $15·80, 9 at $16 = $144, 10 at $15·80 = $158
Actual Cost $ $142·20
3 19 at $3·60, 19 at $4 = $76, 20 at $3·60 = $72
Actual Cost $68·40
4 11 at $22·50, 11 at $22 = $242, 10 at $22·50 = $225
Actual Cost $ 247·50
5 12 at $14·35, 12 at $15 = $180, 10 at $14·35 = $143·50
Actual Cost $172·20
6 28 at $16·25, 30 at $15 = $450, 30 at $16·25 = $487·50
Actual Cost $455

Quick Check

1 $153·50
2 $271·50
3 $460·50
4 $173·60
5 $265·50
6 $280
7 $320
8 $660
9 $880
10 $760

Rules for Divisibility

		Divisible by 2	Divisible by 3	Divisible by 5	Divisible by 9
1	6	✔	✔		
2	30	✔	✔	✔	
3	27		✔		✔
4	36	✔	✔		✔
5	55			✔	
6	49				
7	56	✔			
8	54	✔	✔		✔
9	42	✔	✔		
10	75		✔	✔	

UNIT 26 STATISTICS & PROBABILITY

How Many People in Each State?

1 NSW
2 SA
3 651 769
4 1 691 292
5 21 243 760

UNIT 26 MEASUREMENT & SPACE

Timed Presentations

1 Lasted 10 minutes, 3 minutes shorter.
2 Lasted 9 minutes, 2 minutes shorter.
3 Lasted 14 minutes, 7 minutes shorter.
4 2:46 and 30 seconds
5 Lasted 4·5 minutes, 2·5 minutes longer.

UNIT 26 NAPLAN*-STYLE QUESTIONS

1 $7·40
2 4:35
3 2 h 30 min

UNIT 26 PROBLEM OF THE WEEK

Harshad Numbers

0	**1**	**2**	**3**	**4**	**5**	**6**	**7**	**8**	**9**
10	11	**12**	13	14	15	16	17	**18**	19
20	**21**	22	23	**24**	25	26	**27**	28	29
30	31	32	33	34	35	**36**	37	38	39
40	41	**42**	43	44	**45**	46	47	**48**	49
50	51	52	53	**54**	55	56	57	58	59
60	61	62	**63**	64	65	66	67	68	69
70	71	**72**	73	74	75	76	77	78	79
80	**81**	82	83	**84**	85	86	87	88	89
90	91	92	93	94	95	96	97	98	99

3 The pattern is symmetrical about the main diagonal.

UNIT 27 NUMBER & ALGEBRA

Pattern Rules

1 2, 5, 3, 6, 4, 7, 5. The 9th number will be even.
2 3, 13, 10, 20, 17, 27, 24. The 9th number will be odd.
3 33, 30, 28, 25, 23, 20, 18. The 9th number will be odd.
4 8, 16, 20, 40, 44, 88, 92. The 9th number will be even.
5 = A2 + 2
= A3 * 5
6 = A2 – 3
= A3 * 6
7 = A2 / 2
= A3 + 6
8 = A2 * 6
= A3 + 4

Quick Check

1 + 3, + 2
2 + 6, – 2
3 × 3, – 6
4 – 2, + 10
5 + 1, × 2
6 × 10, + 100
7 + 4, × 3
8 + 10, – 3
9 – 6 + 10
10 × 10, – 1

Word Problems

1 7 × $5 × 2 = $70
2 $36 – 5 × $3·50 + 5 × $10 = $68·50
3 $40 – $12 – $15 = $13
4 $496 – $96 ÷ 40 = 10 weeks
5 $6·50 × 5 × 10 = $325

UNIT 27 STATISTICS & PROBABILITY

Our Pets

1 Dog
2 12
3 Yes
4 6
5 $\frac{3}{4}$ of the class

UNIT 27 MEASUREMENT & SPACE

Treasure Hunt

1–5

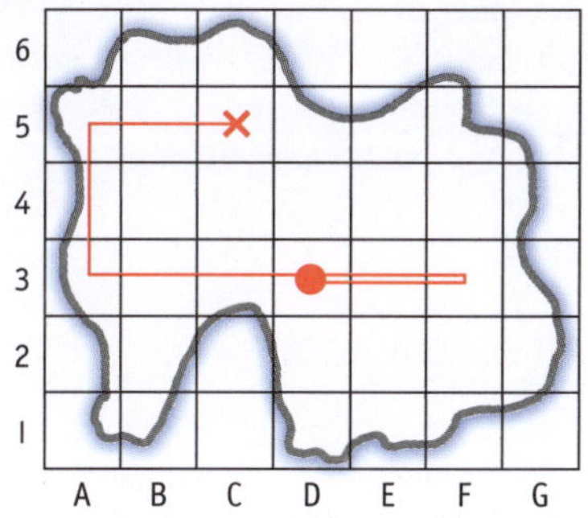

6 (C, 5)

UNIT 27 NAPLAN*-STYLE QUESTIONS

1 65
2 $6
3 (D, 5)

UNIT 27 PROBLEM OF THE WEEK

Button Patterns

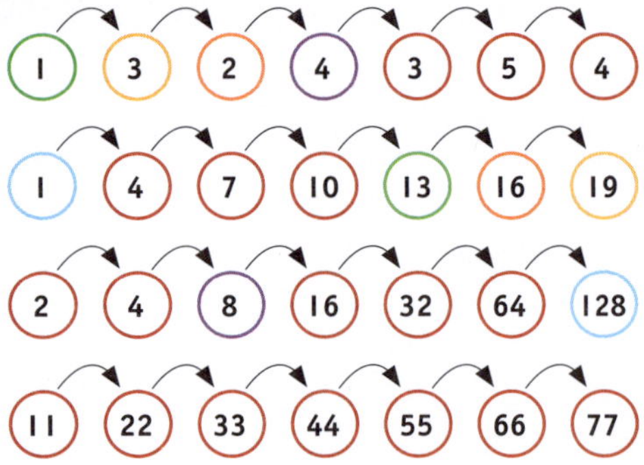

UNIT 28 NUMBER & ALGEBRA

Finding Unknowns in Equivalent Number Sentences

1 36 + 19 = 29 + 26
2 17 + 85 = 45 + 57
3 36 + 16 = 48 + 4
4 165 + 22 = 179 + 8
5 27 − 16 = 38 − 27
6 58 − 5 = 67 − 14
7 103 − 14 = 188 − 99
8 143 − 44 = 103 − 4
9 164 + 6 = 201 − 31
10 85 + 55 = 65 + 75
11 150
12 12
13 10
14 21
15 4

Quick Check

1 38 + 18 = 56
2 59 + 24 = 83
3 76 − 28 = 48
4 83 − 41 = 42
5 76 + 29 = 105
6 26 + 26 + 48 = 100
7 38 + 54 − 32 = 60
8 37 − 14 + 35 = 58
9 29 + 35 − 15 = 49
10 27 − 2 + 16 = 41

Make a Total Even, Make a Total Odd

Open ended

UNIT 28 STATISTICS & PROBABILITY

Preparing for Camp

1 Pineapple, because it got the least votes.
2 Orange juice
3 All four chose apple juice.
4 Do you have allergies?
5 Which juice would you not drink?

UNIT 28 MEASUREMENT & SPACE

Draw a House

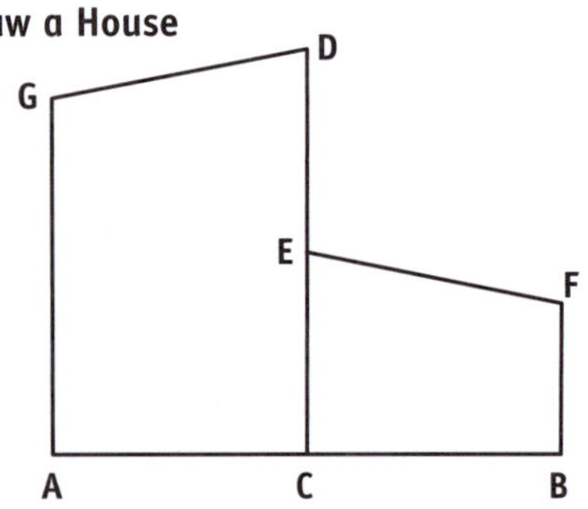

UNIT 28 NAPLAN*-STYLE QUESTIONS

1 36 + 10 = 59 − 13
2 The fourth line
3 9

UNIT 28 PROBLEM OF THE WEEK

Making 20

These combinations of 4 and 5 with + and − use at most 4 numbers. These are the shortest possible expressions for each of the numbers 1–20. A few of the expressions are hard to make so a comment has been added to show how the expression can be found.

1 5 − 4
2 5 + 5 − 4 − 4
(2 is 1 more than 1)
3 4 + 4 − 5
(3 is one less than 4)
4 4
5 5
6 5 + 5 − 4
(6 is one more than 5)
7 4 + 4 + 4 − 5
(7 is one less than 8)
8 4 + 4
9 5 + 4
10 5 + 5
11 5 + 5 + 5 − 4
(11 is one more than 10)
12 4 + 4 + 4
13 5 + 4 + 4
14 5 + 5 + 4
15 5 + 5 + 5
16 4 + 4 + 4 + 4
17 5 + 4 + 4 + 4
18 5 + 5 + 4 + 4
19 5 + 5 + 5 + 4 (one less than 20 … and so on)
20 5 + 5 + 5 + 5 (20 is the best place to start)

UNIT 29 NUMBER & ALGEBRA

The Safest Route

1 36·5 × 2 = 73, 73 × 10 = 730
2–4 A typical safe route would be 6·4 × 10 = 64, 64 × 100 = 6400, 6400 × 10 + 500 = 64 500

Quick Check

1 400
2 3·6 × 10 = 36
36 × 100 = 3600
3 5·7 + 0·3 = 6
6 × 100 = 600
4 5·8 − 0·4 = 5·4
5·4 × 100 = 540
5 6·8 + 0·2 = 7
7 × 1000 = 7000
6 4·9 + 0·1 = 5
5 × 1000 = 5000
7 6·05 + 0·05 = 6·1
6·1 × 10 = 61
8 6·25 + 0·75 = 7
7 × 10 = 70
9 7·75 − 0·25 = 7·5
7·5 × 10 = 75
10 28·6 × 100 = 2860

Salad Plates

1 4 radishes
2 $\frac{1}{2}$ avocado
3 $\frac{3}{4}$ tomato
4 3 spring onions
5 6 cheese cubes

UNIT 29 STATISTICS & PROBABILITY

Pull out a Red Counter

1 Likely
2 Impossible
3 Unlikely
4 Unlikely
5 Likely

UNIT 29 MEASUREMENT & SPACE

Morning or Afternoon

1 pm
2 am
3 am
4 pm
5 pm
6 12 hours
7 14 hours
8 4 hours
9 10 hours
10 18 hours

UNIT 29 NAPLAN*-STYLE QUESTIONS

1 3·1
2 2:20 pm
3 The fourth spinner

UNIT 29 PROBLEM OF THE WEEK

Just by Chance

Examples are:

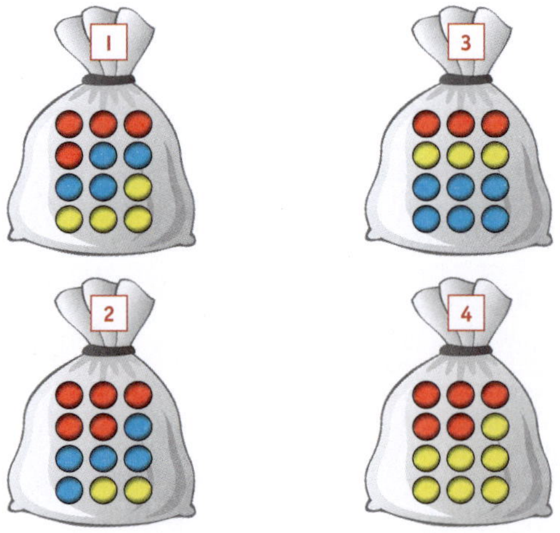

ANSWERS

TARGETING MATHS HOMEWORK: YEAR 4 © PASCAL PRESS ISBN: 9781925726572

UNIT 30 NUMBER & ALGEBRA

Fractions

Some have more than one possible answer.

1. The $\frac{1}{2}$ cup
2. The $\frac{1}{3}$ cup
3. The $\frac{1}{2}$ cup and the $\frac{1}{4}$ cup
4. The $\frac{1}{8}$ cup
5. Use $4 \times \frac{1}{8}$ cups and the $\frac{1}{2}$ cup.
6. Use $2 \times \frac{1}{8}$ cups, the $\frac{1}{4}$ cup and the $\frac{1}{2}$ cup.
7. Use $6 \times \frac{1}{8}$ cups and $1 \times \frac{1}{4}$ cup.
8. Use $2 \times \frac{1}{8}$ cups, $2 \times \frac{1}{4}$ cups and $1 \times \frac{1}{2}$ cup.
9. Use $6 \times \frac{1}{4}$ cups.
10. Use $2 \times \frac{1}{8}$ cups and $2 \times \frac{1}{4}$ cups.

Quick Check

1. 1 cup
2. $\frac{3}{4}$ cup
3. $\frac{2}{8}$ cup or $\frac{1}{4}$ cup
4. $\frac{2}{3}$ cup
5. $\frac{3}{8}$ cup
6. 6
7. 4
8. 3
9. More
10. 6

Finding Halves and Quarters of Money Amounts

1. half of \$7 = \$3·50
2. half of \$3·60 = \$1·80
3. half of \$27 = \$13·50
4. half of \$105 = \$52·50
5. quarter of \$8 = \$2
6. quarter of \$9 = \$2·25
7. quarter of \$6·80 = \$1·70
8. quarter of \$17·60 = \$4·40
9. quarter of \$120 = \$30
10. quarter of \$24·40 = \$6·10

UNIT 30 STATISTICS & PROBABILITY

Cheaper by the Packet

1. \$10·80
2. C4
3. 3 carton of OJ
4. \$1·40
5. \$1·80

UNIT 30 MEASUREMENT & SPACE

Symmetrical Shapes

Examples are:

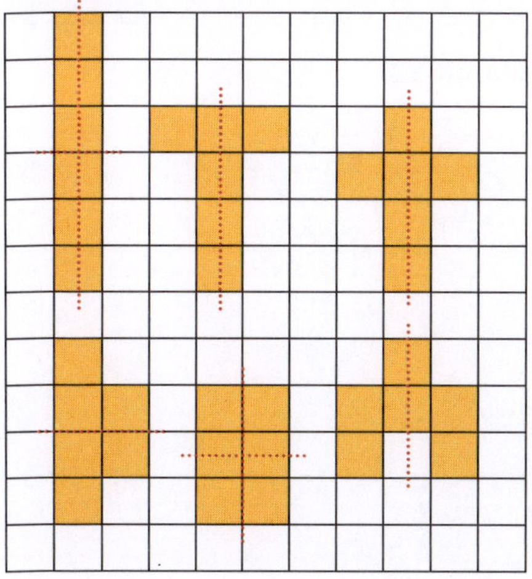

UNIT 30 NAPLAN*-STYLE QUESTIONS

1. The first one
2. \$12·50
3. Paulo

UNIT 30 PROBLEM OF THE WEEK

The Pizza Party

1. One-quarter of a pizza was left for Zoe.
2. Jo

UNIT 31 NUMBER & ALGEBRA

Comparing Fractions

1. $\frac{3}{4}$ is larger
2. $\frac{6}{8}$ is larger
3. Same
4. $\frac{7}{8}$ is larger
5. $\frac{5}{6}$ is larger

Quick Check

1. $\frac{1}{5}$
2. $\frac{1}{3}$
3. Same
4. $\frac{1}{2}$
5. Same
6. $\frac{1}{4}$
7. $\frac{3}{4}$
8. $\frac{1}{2}$
9. $\frac{1}{2}$
10. $\frac{7}{8}$

Fractions and Division

1. $\frac{1}{4}$ of 8 = 2, 8 ÷ 4 = 2
2. $\frac{1}{5}$ of 10 = 2, 10 ÷ 5 = 2
3. $\frac{1}{6}$ of 12 = 2, 12 ÷ 6 = 2
4. $\frac{1}{3}$ of 12 = 4, 12 ÷ 3 = 4
5. $\frac{1}{8}$ of 16 = 2, 16 ÷ 8 = 2

UNIT 31 STATISTICS & PROBABILITY

The Spellathon Results

1. 40–50
   50–60
   60–70
   70–80
   80–90
   90–100
2. 100 students
3. 25 students
4. 65 students

UNIT 31 MEASUREMENT & SPACE

The Garden Design

1. 22 m
2. 7 m²
3. 8 m²
4. 8 m
5. 8 m²
6. 2 m²

UNIT 31 NAPLAN*-STYLE QUESTIONS

1. The fourth one
2. D
3. 4

UNIT 31 PROBLEM OF THE WEEK

Counters by Colours

1. Claire had 3 red, 6 yellow and 9 green counters.
2. Jake had 10 red, 5 yellow and 5 green counters.
3. Chen had 12 red, 3 yellow, 6 green and 9 black counters.
4. Open ended

UNIT 32 NUMBER & ALGEBRA

Money Amounts

1. 60 ÷ 20 = 3, which makes \$3
2. 120 ÷ 10 = 12, which makes \$12
3. 12 ÷ 2 = 6, which makes \$6
4. 80 ÷ 5 = 16, which makes \$16
5. 80 ÷ 20 = 4, which makes \$4
6. 140 ÷ 20 = 7 which makes \$7
7. 50 ÷ 2 = 25 which makes \$25
8. 220 ÷ 10 = 22 which makes \$22

Quick Check

1. \$2
2. \$2·80
3. \$11
4. \$20
5. \$60
6. \$15
7. \$1·05
8. \$1·10
9. \$4·20
10. \$17

Money Pattern Rules

1. \$0·05, \$0·10, \$0·20, \$0·40, \$0·80, \$1·60, \$3·20, \$6·40, \$12·80, \$25·60
2. \$6·50, \$6·70, \$6·90, \$7·10, \$7·30, \$7·50, \$7·70, \$7·90, \$8·10, \$8·30
3. \$2·05, \$2·55, \$3·05, \$3·55, \$4·05, \$4·55, \$5·05, \$5·55, \$6·05, \$6·55
4. \$0·15, \$0·25, \$0·45, \$0·85, \$1·65, \$3·25, \$6·45, \$12·85, \$25·65, \$51·25
5. \$0·55, \$0·60, \$0·70, \$0·90, \$1·40, \$1·45, \$1·55, \$1·75, \$2·25, \$2·30
6. Largest is pattern 4.

UNIT 32 STATISTICS & PROBABILITY

Pick out 40c

Mostly open ended, except that the chance of making more than 30c is 'Likely'.

UNIT 32 MEASUREMENT & SPACE

Shape Creatures

1

	Shape A	Shape B	Shape C
Isosceles triangle	2	2	
Square	2	2	
Rectangle		1	1
Right-angle triangle	2	2	5

2 4 squares
3 7 squares
4 5·5 squares

UNIT 32 NAPLAN*-STYLE QUESTIONS

1 61 balloons
2 1
3 square

UNIT 32 PROBLEM OF THE WEEK

Coins in the Purse

- Claire: If all my coins were 5c coins, I would have $2·20 and I would be $1·80 short. Every time I change a 5c for a 20c, I get 15c more. If I change 12 of the 5c into 20c I would have an extra $1·80. For me to have $4 with 44 coins, I need 32 × 5c coins to make $1·60 and 12 × 20c coins to make $2·40. That would be $4 in all.
- Jo: If all my coins were 10c coins, I would have $4·40 and I would be $3·60 short. Every time I change a 10c for a 50c, I get 40c more. If I change 9 of the 10c into 50c I have an extra $3·60. For me to have $8 with 44 coins, I need 35 × 10c coins to make $3·50 and 9 × 50c coins to make $4·50. That would be $8 in all.
- Claire would have 40 × 5c coins to make $2 and 20 × 20c coins to make $4. That would be $6 in all.
- Jo would have 45 × 10c coins to make $4·50 and 15 × 50c coins to make $7·50. That would be $12 in all.

TERM 4 REVIEW

Review: Unit 25 and Unit 26

1 30 + 20 = 50, 40 + 30 = 70.
The range is between 50 and 70.
2 40 + 50 = 90, 50 + 60 = 110.
The range is between 90 and 110.
3 70 + 80 = 150, 80 + 90 = 170.
The range is between 150 and 170.
4 190 + 30 = 220, 200 + 40 = 240.
The range is between 220 and 240.
5 Open ended
6 $4 + $2 + $4 = $10. A safe estimate is $10.
7 $17 + $4 + $12 = $33. A safe estimate is $33.
8 8 at $2 = $16, 10 at $1·65 = $16·50
9 9 at $3 = $27, 10 at $2·20 = $22
10 27 and 45
11 30 and 15
12 45, 90
13 1:15 pm
14 5 h 30 min

Review: Unit 27 and Unit 28

1 +4 −3 +4 −3 +4 −3
6 10 7 11 8 12 9
2 ×2 +1 ×2 +1 ×2 +1
3 6 7 14 15 30 31
3 2 4 3 5 4 +2 and −1
4 5 10 9 18 17 34 ×2 and −1
5 40 − 4 × 7 = 12
Jake has $12 left.
6 30 × 5·5 = 165
Claire earned $165 in June.
7 $\frac{1}{4}$
8 $\frac{1}{2}$
9 27 + 35 = 37 + 25
10 56 − 16 = 76 − 36
11 59 − 24 = 16 + 9
12 68 − 43 = 40 − 15
13 55 and 63
14 26 and 35 or 35 and 22
15 AB = 3 cm, AC = 4 cm, BC = 5 cm

Review: Unit 29 and Unit 30

1 From 3·6 to 40
3·6 + 0·4 = 4
4 × 10 = 40
2 From 6·7 to 700
6·7 + 0·3 = 7
7 × 100 = 700
3 2 olives
$\frac{1}{2}$ a slice
3 salami
4 Unlikely
5 5 hours
6 $\frac{3}{4}$
7 $\frac{1}{2}$
8 $17·50
9 $2·75
10 $3·85
11 $15·50
12 95 cents
13 Sample answers:

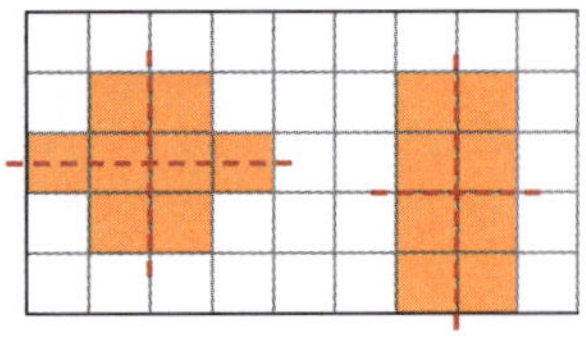

Review: Unit 31 and Unit 32

1 $\frac{5}{6}$
2 $\frac{3}{5}$
3 $\frac{1}{4}$ of 12 = 3, 12 ÷ 4 = 3
4 81–90
5 51–60 and 61–70
6 $\frac{1}{3}$
7 20 m
8 24 m^2
9 $2·50
10 $2·60
11 $15·50
12 $10·10
13 $6\frac{1}{2}$ squares

TARGETING MATHS HOMEWORK: YEAR 4 © PASCAL PRESS ISBN: 9781925726572